THE SUBJECTIVE SIDE OF SCIENCE

A Philosophical Inquiry into the Psychology
of the Apollo Moon Scientists

THE SUBJECTIVE SIDE
OF
SCIENCE

A Philosophical Inquiry into the Psychology of the Apollo Moon Scientists

Ian I. Mitroff

Graduate School of Business,
Interdisciplinary Doctoral Program in Information Science
and The Philosophy of Science Center
University of Pittsburgh

 Elsevier

ELSEVIER SCIENTIFIC PUBLISHING COMPANY
335 JAN VAN GALENSTRAAT
P.O. BOX 211, AMSTERDAM THE NETHERLANDS

AMERICAN ELSEVIER PUBLISHING COMPANY, INC.
52 VANDERBILT AVENUE
NEW YORK, NEW YORK 10017

Library of Congress Card Number: 74-77584

ISBN 0-444-41233-6

Contents

To my teachers
C. West Churchman
and
Thomas A. Cowan

List of Figures

List of Tables

Foreword

"We glory in our tribulations, knowing that tribulation produces patience, and patience experience, and experience hope; and we are not ashamed of hope." St. Paul, *Romans*, 5

We have come to use the label *scientific method* to express an acceptable way of developing knowledge of the world in which we humans live. Acceptable to whom? The answer to this very sensible question can be regarded as the essence of the scientific method. In the sixteenth and seventeenth centuries, philosophers like Descartes and Locke, different as they were in their world views, forged a common basis for the criterion of acceptability. It was based on the speculation that all humans are endowed with a common capability of reasoning and observing. Of course, many other psychological functions interfere with these two capabilities, e.g., anger, fear, pride, and conformity. But, so ran the speculation, there are explicit ways of stripping away these personal and social contaminations, so that common reason and common observation operate more or less in a pure state. When this purification occurs, the resulting method, directed at a particular target question, will yield results acceptable to all who have attained the pure state, or a reasonable approximation thereof.

Over the centuries that followed the monumental work of the rationalists and empiricists, a great deal has been done to refine, modify, and understand this speculation about scientific method. Technically, it was essential to create a methodology of controlled observation, which we call measurement, and to develop a theory of errors to cope with the fact that no two measures can be expected to agree thoroughly. So too, the laws of reasoning needed far better refinement than was available in the Renaissance. But the philosophical speculation also required reexamination, beginning with Kant's astonishing conclusion that reason and observation are nonseparable functions, so that, for example, the regularity we supposedly find in nature is, in fact, put there by the need to reason in order to observe.

It is probably safe to say that the scientific community accepted the Descartes—Locke speculation up to a few decades ago, even though it became more and more obvious that purifying the investigator was often an enormously difficult task; professors contaminate their students, great and forceful scientists contaminate everyone, the reward system contaminates wholly, and so on.

However, we now seem to be entering an age in which many scientists — historians and philosophers of science included — are beginning to suspect that the great speculation is not valid: It is impossible to purify ourselves of "contamination"; or, there is no common reason and observation to attain which makes us objective.

I think the first sciences to realize this were the applied social sciences, and the roots of their doubt may very well reach back into the nineteenth century. The applied social scientist, if he is general in his thinking, comes to realize that he is fundamentally a social being. Any investigation he conducts of his biases must be based in part on the kind of social being he is. For example, many investigators must operate within what Kuhn calls the paradigm of their discipline, so that an economist or a behavioral scientist must each look at the social world differently because of their educational background. Or, it is all too natural for Forrester* to liken the social world to a machine, and Beer** to liken it to a brain. We have no idea at all how these applied social scientists could decontaminate their method of investigation, and indeed we suspect that this is not the problem. The problem is how or whether humans can learn about society despite their very deep differences in world views. This dialectical speculation, namely, that learning takes place in the context of disagreement rather than agreement, is as old as the ancient Greeks; Socrates used it as a pedagogical method. Hegel seems to be the most prominent espouser of the speculation prior to our century. But none of Hegel's thought really influenced the scientific community, not even in the USSR, which on paper is committed to one form of dialectics. But many of us in applied social science, including the author of this book, have come to realize that the main intellectual business we are in is

---

* See for example, Meadows, D.H. et al., *Limits to Growth* (New York: Universe, 1972).
** Beer, Stafford, *The Brain of the Firm: A Development in Management Cybernetics* (New York: Herder & Herder, 1972).

dialectical inquiry, in which to be objective means to view the object and its world in dramatically different ways.

One might be inclined to say this proves that applied social science is not a science, were it not for a third speculation about science and its methods which has been developing outside the scientific community. This is the speculation that science is itself a social activity and that society pays a price for the benefits of scientific investigation. Just how much the price should be depends on how effectively the investigation is carried out and what its contribution is to society. Hence — so runs this speculation — scientific method is an inquiring system which produces socially beneficial knowledge in as effective a manner as possible. This is surely the speculation that is growing in the agencies of the largest research supporter in the world, the government of the United States. It may appear to be a very crass speculation to those who love pure science for its own sake, but it has a very strong ethical justification: Why should citizens pay for research which they cannot possibly understand and which is not likely to benefit them? If scientists take this challenge seriously, they land in a rich dilemma. Either the assessment of a research project is to be accomplished by scientific method or it is not. If it is, then the assessment is a project of applied social science, and hence cannot be done by traditional scientific methods. If it is not, then the basic principles of science which justify its existence are nonscientific. Again, many of us in applied social science, including the author, accept the challenge and are wishing to land on the first horn of the dilemma, in which strong disagreement is the essence of inquiry.

The story of the three speculations about scientific method brings us to the main topic of this book. It has become rather old hat to say that the causes of disagreements among scientists are social in kind; it is rather new hat to say — as does this book — that they are also psychological and that this is all to the good! The scientists who disagree so strongly in these pages do *not* thereby violate the principles of sound inquiry. There are very sound psychological reasons why an inquirer *should* hold onto his convictions even though his colleagues believe the evidence is against him. This point is entirely missed in the traditional speculation about scientific method, where we are supposed to believe that if a theory implies a result which observation refutes, then the theory must be abandoned. The psychological counterpart of this myth says that if I am totally committed psychologically to a conviction, and I seem to observe something that runs counter to the conviction, I should abandon the conviction. If people behaved in accordance with this psychological principle they would become insane. The Church, after all, did have a point in its case against Galileo, if the people of that time were totally unprepared psychologically to shift from an earth-as-universe-center archetype to earth-as-insignificant-rock. Truth is as much psychological as it is logical. The coming of age of science may very well be the scientist's realization of the place of inquiry in the total psyche of the individual.

The heroes of this book are human, above all, and therefore, by facing their tribulations humanly, have laid the foundations for being scientists.

C. West Churchman
University of California
Berkeley

Preface

This is a book about the nature of science and about how to study science. As I argue throughout, these two major themes cannot be talked about in isolation. A theory about the nature of science is also a theory of how to study science. Above all, this book is an essay about how scientists often have exceedingly good reasons for behaving irrationally. It is a direct challenge to some of our current conceptions of science. It is a foray into the region where the boundaries between the philosophy, sociology, and psychology of science become almost indistinguishable.

The aims of the book are basically twofold: (1) to lay the foundation for an improved model for the practice of science that is founded on a critical appraisal of the actual behavior of scientists, and (2) to ask what a science would look like that better understood itself.

The number of people who have been instrumental in the initiation of this study and in the writing of this book is too large for me to mention every one of them by name. Nevertheless I would like to gratefully acknowledge their support and assistance. I would also like to pay special thanks to the following persons: to Richard E. Stephens and Eugene Marr, of the Office of University Affairs, NASA, for their role in approving the grant application which allowed me to do this study; to Frank Hansing, also of the Office of University Affairs, NASA, and Homer E. Newell, Associate Administrator, NASA, for their role in the initial and continued funding of this study. Richard E. Stephens in particular deserves a special word of mention and thanks. He not only supported this study from the very beginning but he has also helped to see it through every one of its subsequent phases.

The following persons read parts or all of the manuscript and offered valuable criticisms or comments on it. Although I may not have incorporated all of their suggestions for improvement, the book is better as a result of their criticisms and comments: David Bakan, Carl Beck, Richard Cottam, Paul Diesing, William C. Frederick, Michael Gold, Jonathan Hodge, Burkart Holzner, Allen Kent, Laurens Laudan, Paul F. Lazarsfeld, Michael Martin, Alex Michalos, Robert Perloff, Derek de Solla Price, Jerome Ravetz, Nicholas Rescher, Loyd Swenson, Walter Weyrauch, William F. Whyte, Richard Willis, and James A. Wilson. In particular I cannot thank Alex Michalos and Nicholas Rescher enough for their many efforts in my behalf. I have greatly profited from discussions with Thomas Kuhn and Robert Merton. I would also like to thank Nicholas Wilson of the Clarendon Press, Sanford Thatcher of the Princeton University Press, and Frederick Hetzel of the University of Pittsburgh Press for their many helpful comments and criticisms.

Special thanks are due to James Williams, John Nelson, and Debby Drake for their help with the many computer calculations. I wish in this regard also to express my strong appreciation to Randy Chapman who helped with the Multidimensional Scaling Analyses of Chapter Four. I cannot thank enough my research assistant, Inez Fitzgerald, without whose help I could never have analyzed the seemingly endless rounds of transcript material and questionnaire data. Dianne Krooth played an important role in helping me with part of the fourth and final round of interview questions.

Rose Fazio performed the fundamental and enormous task of transcribing the tape-recorded interviews. Gerry Lee and Grace Bauman more than competently typed the entire manuscript. I gratefully acknowledge their tireless efforts.

No book is ever the product of a single mind. It is always the product of many minds and many influences. In my own case, the most powerful influences were those of my teachers, C. West Churchman and Thomas A. Cowan. They have touched nearly every aspect of my life. Although I did not study directly under him, I also consider Russell L. Ackoff one of my teachers because I have learned so much from his writings and in personal conversations with him. I thus wish to acknowledge here his enormous influence on my intellectual life.

In many senses this book is the direct outcome of a philosophy of science seminar that took place over the course of three years during the latter phases of my graduate study at Berkeley. The seminar was conducted by Churchman and Cowan, whose influence I have already acknowledged. Fred Betz, Douglas Day, Michael Echols, and Richard O. Mason were fellow students in that seminar. They were as much my teachers as my fellow students. They remain close friends, invaluable colleagues and critics. I greatly prize their warm friendship. In this regard, I cannot say enough to acknowledge my debts, both personal and intellectual, to Vaughn Blankenship. Vaughn and I used more than a goodly number of telephone message units every week between Pittsburgh, Buffalo, and Washington, D.C., discussing nearly every aspect of this study. His friendship and wise advice helped me over many hurdles.

To my own students, John Nelson, Chai Kim, Paul Peters, Eugene Rathswohl, and Inez Fitzgerald, I am grateful because they have taught me more than I have taught them.

Finally to the best and dearest friends I have, my wife Donna and my daughter Dana, these few words are but the tiniest expressions of the love I feel for them both. Without them none of this would have been possible; without them none of it would have mattered. They know a lot about the nature of the moon without having to think about it at all.

This work was partially supported under NASA Grant NGL 39011080. The opinions, interpretations, and conclusions expressed in this book are solely those of the author and do not in any way represent the opinions, official or unofficial, of any governmental of private agency.

Ian I. Mitroff
Pittsburgh, Pennsylvania
March 1974

Everyone is a moon, and has a dark side which he never shows to anybody.

Mark Twain

PART I

Theoretical Beginnings

Chapter One

The Moon is a Social Science Problem

"Nature is not organized in the same way that universities are."

Russell Ackoff [13, p. B-127]

"The autonomy of academic departments can be respected only at the peril of insuring our ignorance."

Abram Kardiner [220, p. 90]

"It has been put to me that one should in fact distinguish carefully between Science as a body of knowledge, Science as what scientists do, and Science as a social institution. This is precisely the sort of distinction that one must *not* make... Before one can distinguish and discuss separately the philosophical, psychological or sociological dimension of Science, one must somehow have succeeded in characterizing it as a whole."

John Ziman [486, pp. 11—12]

This is a book about how science actually gets done. As with most areas of human conduct, the actual picture is very different from the stereotyped image that abounds in our popular literature and in too much of our academic literature. This is as much a book about how to study science as about the nature of science. It is a critical study of many of the cherished assumptions that underlie intellectual life in general and the sciences in particular.

The context of the study is the background to one of man's greatest adventures, the landing of men on the moon and their safe return to earth. The subjects were more than 40 of the scientists who participated in the Apollo lunar missions. The basic purpose of the study was to test specific contemporary and critical propositions in the sociology, psychology, and philosophy of science by means of a combined philosophical and social science investigation into the attitudes, beliefs, and scientific practices of an interesting, specific group of scientists. Each scientist in the study was intensively interviewed on four separate occasions during a span of three years. As a rule, the scientists were interviewed during the period just after the completion of one Apollo mission and before the start of another — between Apollo 11 and 12, 12 and 14, 14 and 15, and 15 and 16. Each respondent was interviewed four times in order to determine whether the beliefs of the scientists were changing in response to the scientific data brought back from successive Apollo missions.

The purposes of this chapter are: (1) to convey the personal and intellectual stirrings out of which this study was conceived and grew, and (2) to introduce some of the general philosophical and social science themes, as well as specific propositions, to which this book is addressed. I will come back to these themes and propositions in more detail after I have presented and assessed critically the attitudinal data which were gathered from the scientists and which bear on the issues we will be examining.

The Beginnings: "Marvelous Little Moon Rocks"[1]

Robert Vas Dias wrote: "The premise... is that outer space is as much a territory of the mind as it is a physical concept" [450, p. xxxix]. On July 20, 1969, when Buzz Aldrin and Neil Armstrong stepped on the moon, mankind was, as perhaps never before, made dramatically aware of its accomplishments in science and technology. Behind this incredible feat are almost four centuries of sustained effort in the sciences. One is tempted to say that the landing of men on the moon is one of the direct by-products of the cumulative growth and sustained advance of progress in the sciences.[2]

As incredible as this feat was, an even greater drama (at least for those of us who are studying the sociology and psychology of scientists) was forming behind the scenes. In many respects, the most fantastic journey was not the one that transported and landed men on the moon but the journey which began with the return of the moon rocks to the scientists here on earth. The real drama, the real cosmic play, concerns how those rocks have affected our understanding of the moon, of science, and ultimately of man. The technological and political goal of landing men on the moon pales in comparison to the scientific goal of understanding the nature of the moon and the goals of understanding the nature of

4

science and of man. It is easier to land men on the moon than it is to understand the nature of the moon. The reason, as many of our best social critics [99, 270, 321—324] have been quick to point out, is that going to the moon does not require any fundamental change in our ability to understand ourselves. Indeed, some would go so far as to contend that this very lack of any increase in our self-understanding is almost a necessary precondition for undertaking such costly and childish ventures as moon trips. The goal of understanding the nature of the moon is a very different one. This book will argue that if we are to understand the nature of the moon we must achieve a far greater understanding of our own nature. To be more specific, we must achieve a far greater degree and kind of insight into one of the grandest and most mysterious — and in many ways most frightening — of all of man's creations, science, before we can even pretend to understand an object like the moon.

If these thoughts seem strange, perhaps it is because we are not as reflective as we would like to believe we are. Accustomed and conditioned as we are by the thoughts of our time, we think it only natural to keep the technology of space ships and the physics of moon rocks strictly apart from the psychology and sociology of earth men. We really think we can increase our understanding of the moon as an inanimate, natural object without simultaneously increasing our knowledge of ourselves as reflective, human observers.

The title of this chapter emphasizes my belief that it is always men who speak for nature in terms of the highly select questions they so carefully and deliberately choose to address to her. Moon rocks do not speak and tell us about the moon; only men who know how to address mother nature properly can tell us about the moon. The characteristics of men, as expressed in the ways men go about knowing the physical world and the properties of rocks, are often so intertwined that it becomes extremely difficult, if not impossible, to say where the purely physical leaves off and where the psychological begins. Our knowledge of the so-called purely physical properties of the world contains more than an insignificant trace of the projective side of our psychological nature. It is the special job of the social sciences to study critically how the properties of men affect what they know of the physical world.

These thoughts obviously reflect a good part of the final outcome of this study. Nevertheless, more than just the spirit of these thoughts was going through my head as I, with millions of others, stayed up that night to watch those two men step out onto the moon. As I looked at men walking on the moon that night, the significance of what was unfolding began to make itself clear to me. Rocks from the moon were going to be returned to earth in order to test earth-based scientists' ideas about the moon. Men were either going to have their theories and ideas vindicated or not. Even though I was not then intimately familiar with the competing theories about the nature of the moon, I knew that there were many. Theories about the moon's nature go back as far as we have evidence about man's history; modern scientific ideas go back at least as far as Kant [211]. I also knew, although I was not then intimately familiar with the various proponents of these theories, that they were hotly debating the merits of their respective theories. More than one author had committed himself in print as to what we might expect to find; some had even written of what they hoped we would find. Perhaps the passionate, and often even irrational, adherence to ideas was the norm or distinguishing mark of the creative scientist [234, 458]. If so, this might be especially true for those scientists who are bold, imaginative, and capable enough to propose theories of origin for something so huge and complicated as a whole earth-moon system. If this were indeed more the norm than the widely held and proclaimed image of the scientist as a cold, emotionally disinterested creature guided only by the dictates of impersonal logic and a rational craving for the truth, then one would expect the various proponents of theories not to give in in the face of the first bits of negative evidence. In fact, such men would undoubtedly argue about what constituted negative evidence. Because no data are ever purely for or against any theory, no matter how outrageous that theory may be, would not the proponents of such a theory hold on to their ideas for some time, even in the face of negative evidence? Would they not continue to debate the validity of the data? Would they not argue over whether the data was representative of the moon, or was a biased sample, or was in some strange and unexpected way artifactual? Such thoughts led me to hypothesize that the case of the moon scientists might provide an excellent historical setting to test a number of ideas about the nature of science that had slowly been accumulating in recent years from a variety of separate disciplines [46, 247, 252, 291, 292, 371]. The field of lunar science certainly seemed to be on the verge of a revolutionary chapter in its scientific history, a chapter that should be recorded in some way other than the usual forms of recording scientific events.

Given the initial insight, all sorts of questions began to form. Certainly I would not expect all proponents of a particular theory for the moon's origin to hold on to that theory to the same degree as they faced countervailing evidence. I would expect some degree of persistence from each theorist. Pet ideas are not easily relinquished. Some of the proponents had been arguing their positions for over a decade. One does not live with such global thoughts for that long unless one is committed to them, and with more than just a modicum of strength. But people are too variable to *all* hold on to their ideas to the same degree. Who, then, would shift the most, first? What would be the psychological characteristics or differences between those who shifted first and most easily and those who seemed prepared to resist to the very end? How would those who held on to the very end be regarded? Would this behavior be just par for the course? These were only some of the initial questions. Many other questions developed as the project became better formulated.

Instead of detailing all the questions that developed, it is more important at this point to note that, like many scientific studies, this one did not begin with detailed and clearly formulated hypotheses that could immediately be tested with strict statistical precision. It began with the vague stirrings of intuition that there was something terribly important to ask about the significance of what was taking place on the moon. More than rocks were being unmooned and were to be placed under the microscope. A unique opportunity to study the institution and methods of science was to be afforded. If the various proponents continued to hold on to and defend their theories in the face of the evidence provided by the moon rocks that were brought back, we would have a rare opportunity to study the participants in an on-going scientific controversy. We can only speculate as to how much more we might know about the nature of science if we had been able, for instance, to interview or record systematically the views of both Newton and Leibnitz regarding their conflicting claims to priority in discovering the calculus. It would be a terrible shame if we were to lose another chance to study scientifically the behavior of the participants in an important scientific event and controversy.

Storybook Image of Science

This book is basically critical of a view of science which, for reasons that shall become apparent, I shall deliberately label The Storybook Image of Science.[3] It is a Storybook view because it is the kind of picture that we should expect of children; it is an image of science that is lacking in sophistication or any deep insight into the nature of science. As incredibly naive as this image of science is, it is important to emphasize that I am not talking about or concocting a straw man.[4] There is evidence [468] that some scientists believe in it (we will have occasion to examine why later). It is an image of science that is encountered in textbooks and in popular accounts of science too frequently to be easily dismissed. It would be easier to dismiss the Storybook view outright if it had not been held by so many intelligent and otherwise discerning men of importance and influence. And so an analysis of the widespread currency of the Storybook account is needed.

Although the Storybook image takes many forms and involves many qualities or components, the single quality that one encounters most frequently is emotional neutrality. Because this book is so critical of the Storybook account, it may be helpful to present some short selected portions from at least one version of this account. The reader can then see what I regard as a Storybook view of science. I have chosen the following version of the components of the scientific attitude for various reasons: (1) The publication date (February 1967) indicates the currency and persistency of the Storybook account; (2) the fact that this version appears in a science education publication is an indication of the first image of science that the young student of science is likely to encounter; and finally (3) this version provides a baseline standard, however vulgar, by which we can judge the superiority of supposedly more sophisticated, less vulgar acounts:

> *Willingness to change opinions.* People hate to give up an idea — especially one they thought clever, or sacred, or a part of the general structure of ideas on which their security depends. A scientist feels these pangs, also, but is more willing than most to alter an opinion, once he sees reliable evidence to the contrary, because he knows that, every time he does so, he has learned something. Retaining the old opinions intact is satisfying to the ego, but a sure indication that one has learned nothing.
>
> *Humility.* Most people are extremely arrogant in their opinions. In

any tavern — and even on a good many lecture platforms — one will hear opinions expressed with supreme confidence that could not be tested or proved. A scientist realizes how little is known with any certainty; he commonly looks for little truths that the unscientific would consider not worth the trouble.

Loyalty to truth. A scientist is sometimes subjected to humiliation as his findings shift and invalidate some conclusion to which he has previously committed himself, but his loyalty to truth is such that he would rather cut off his right arm than suppress the new data. The general picture is of lofty and even noble devotion to the facts, however they may affect one personality.

An objective attitude. A scientist has a high regard for facts and tries to behave in accordance with them; while an unscientific person tends to see only the facts he wishes to see and to react emotionally against others.

Suspended judgement. A scientist tries hard not to form an opinion on a given issue until he has investigated it, because it is so hard to give up opinions already formed, and they tend to make us find the facts that support the opinions. This is closely related to a desire to investigate before acting — to get all the relevant facts if immediate action is not necessary. There must be, however, a willingness to act on the best hypothesis that one has time or opportunity to form [120, pp. 23-24].

In virtually every version of the Storybook account. The notion of emotional neutrality or detachment makes its appearance. In some versions, it is even phrased as strongly as the *absence of emotion.* For example, Richard H. Lampkin [253, p. 356] lists a number of attitudes that are supposed to be indicative of the scientist's general "renunciation of emotion":

The scientist deliberately renounces all emotion and desire, except that of accomplishing his twofold purpose.

The scientist should never make pontifical announcements, nor indulge in melodrama.

If the scientist is ignorant on certain points, he must acknowledge his ignorance on those points.

The scientist's acknowledgment of his present ignorance is not a resignation to defeat; it is rather that which leaves open the way for future investigation.

Robert L. Ebel [123, p. 78] lists a number of attitudes under the heading of "readiness to think coldly":

(1) Readiness to be impersonal and disinterested in thinking.

(2) Readiness to be unemotional, dispassionate, and thoroughly self-controlled in thinking.

(3) Readiness to control imagination.

(4) Readiness to avoid being swayed by oratory.

(5) Readiness to avoid being swayed by the mere novelty or sensationalism of an idea.

Whether scientists actually are disinterested, impersonal observers and thinkers or instead are highly committed, often passionate, partisan advocates for a particular point of view (theory, hypothesis) is a question that will occupy us throughout this book. [5] Indeed, the central question with which this study began and is concerned throughout is: Would it be possible to identify and to study those scientists, if any, who exhibited a high degree of prior commitment (that is, before Apollo 11) to particular hypotheses or theories regarding the nature or origin of the moon and who thus showed, as a result, a high degree of reluctance to give up their favorite hypotheses in the face of or in spite of the data subsequently brought back from the moon? In the interviews with the 40 scientists, it was found that a small number of them were overwhelmingly and consistently nominated by their peers as the ones "most likely to hang onto their pet hypotheses 'til death do them part." In later chapters, I shall examine the perceptions of these few key scientists by their peers (the 40 scientists who were interviewed) over the course of the interviews, looking at the implications of those perceptions for the sociology, psychology, and philosophy of science. I show that it is possible to measure systematically and precisely the differences in psychology between (1) those scientists who were judged most likely to become committed to favorite hypotheses and to take strong stands on scientific issues and (2) those who were judged most likely to avoid taking strong stands or not to develop intense commitments.

The differences between these two types of scientists are very striking, and they are also, in statistical terms, highly significant. Furthermore, these differences emerge continually. No matter what was used to measure them, the same differences were obtained repeatedly. The study reveals that there are indeed definitive and very strong systematic differences among various types of scientists. Most significant of all is our finding that it is possible to capture and to measure these psychological differences systematically and precisely. And it is possible to determine the significance of these results for the philosophy of science. The importance of this will be made clear later.

Sociologists View of Science

A major portion of this book is concerned with a critical examination of the notion of *emotional neutrality*. An even greater portion is concerned with a critical examination of the rationality of a larger set of notions. This larger set concerns what may be termed the *norms of science* — the ideal institutional standards considered necessary to insure the rationality and the basic character of scientific knowledge.

The main body of current thinking regarding the norms of science derives from the work of the sociologists of science [33, 423], most notably that of Robert Merton [292, 297—299]. Unfortunately we must face a major problem before we can even discuss the norms: Just getting a statement of the norms is a difficult problem. Although agreement and overlap exist among the various formulations and statements [33, 297, 423, 468], there are also crucial areas of disagreement about intent, meaning, and terminology. A second complication is that the thoughts of the most influential worker in the area, Robert Merton, have undergone several formulations and are scattered in several places in the literature. For example, Merton's original 1949 formulation of the norms [297] reads, in my view, substantially like the Storybook version of science. In the 1949 version, the norms are static and overly one-sided. They express a limited and conventional conception of scientific rationality. Later versions [291, 292, 298] are dynamic and dialectical in formulation; rather than science being governed by a single, overly rational set of norms, there is instead a fundamental tug-of-war between two sets of norms that differ fundamentally in character and that capture both the deep rational and the irrational aspects of scientific life. Nevertheless, as far as I can tell, Merton's original (1949) formulation of these norms remains unchanged in the latest edition (1968) of his important book, *Social Theory and Social Structure* [297].

For many reasons, we will be concerned with the following statement of these norms [468]. The basic reasons are (1) that the following set of norms is larger in number than the set originally proposed by Merton, and in this sense, permits of a more refined critique, and (2) that the set below almost purposefully confuses the distinction between norms as ideal *institutional* standards that have nothing to do per se with the behavior of particular individuals and norms as ideal attributes or properties of *individuals*. The following formulation thus raises explicitly the question as to whether, in formulating a normative set of standards for science, the two levels (the sociological and the psychological) *should* be kept apart, even doubting that the two *can* be separated in theory, let alone in practice [16, 72, 77, 79]. One of the purposes of this book is to challenge the arguments of those who have attempted

11

to make the separation [294, 297]. The purpose of that challenge is to show that we hamper our knowledge of science if we persist in drawing the boundaries between disciplines too sharply. Briefly, here is the set of norms that will be considered [468, p. 54; emphasis added]:

Faith in rationality.

Emotional neutrality (as an instrumental condition for the achievement of rationality).

Universalism: In science all men have morally equal claims to the discovery and possession of rational knowledge.

Individualism (which expresses itself, in science particularly, as antiauthoritarianism).

Community: Private property rights are reduced to credit for priority of discovery; secrecy thus becomes an immoral act.

Disinterestedness: Men are expected to achieve their self-interest in work satisfaction and prestige through serving the community interest.

Impartiality: A scientist concerns himself only with the production of new knowledge and not with the consequences of its use.

Suspension of judgement: Scientific statements are made only on the basis of conclusive evidence.

Absence of bias: The validity of a scientific statement depends only on the operations by which evidence for it was obtained and not on the person who makes it.

Group loyalty: Production of new knowledge by research is the most important of all activities and is to be supported as such.

Freedom: All restraint or control of scientific investigation is to be resisted.

By focusing the greater part of my discussion of the sociology of science on the norms of science, I do not thereby mean to imply that this is the only important contribution that the sociologists of science have made to our detailed knowledge and understanding of the social structure of science. Quite the contrary. In contradistinction to other philosophers of science, I not only think the sociology of science has already made important substantive contributions [114, 291—298, 396] to our understanding of science but that it is an indispensable component in our ultimate understanding of science. Science may be many things to many men and may be studied and characterized from an almost infinite number of points of view. But I do not see how one fundamental characteristic can be denied, namely, that science is a social institution and hence that one of its most important descriptions will ultimately be sociological and psychological.

I focus almost entirely on the sociologists' characterization of the norms of science, because I regard those norms as the most fundamental and most general contribution to the study of science by the sociology of science. Hence, if the sociologists are wrong regarding these norms, they deserve to be severely challenged and criticized on this most important and fundamental of points. Above all, it is with these most general and fundamental points that this book is concerned.

Because so much of this book is devoted to a critique of these norms, it is important to make some preliminary statements regarding the form of this critique. There are at least three broad criterion areas with regard to which any set of norms can be evaluated. The first area has to do with *the empirical or factual status of the norms*, that is, how well the norms fare as "factually real descriptors" of the actual working "norms-in-use" [219, 413]. The second area has to do with the *regulative or normative status of the norms as the embodiment of ideal standards of rationality.* That is, with regard to the first criterion area above, the norms (e.g., disinterestedness) may fail miserably to describe the day-by-day working rules that scientists actually adhere to in the heat of actual practice. However, this by itself does not mitigate against the status of the norms on other grounds, namely, as reasonable standards of rationality. Merely because a current set of ideal standards fails to be realized does not mitigate the status of the standards as rational ideals that we ought perpetually to aim toward, even if we never succeed in realizing them completely. To rule against norms because they are unobtainable or because they fail to accord with actual practice is, potentially, to confuse the status of empirical propositions or working rules with the status of normative claims and ideals. The two realms of discourse are not necessarily the same. That is, normative claims may be reasonable or unreasonable, but they are not necessarily true or false. Finally, the third area has to do with *the epistemic status of the norms*, i.e., the argumentative and evidential bases that were used to infer the existential status or serious consideration (postulation) of the norms. This area has to do with the intellectual and scholarly operations that were used initially to derive and then subsequently to argue for the continued postulation of the set of norms under consideration.

The bone of contention of this book is not really the factual basis of the norms of science as they are currently conceived, because even the sociologists of science recognize that these norms often fail to describe actual practice. Indeed, more than anyone, Robert Merton has repeatedly and forcefully pointed out that scientists deviate markedly in their behavior patterns from the very norms that he has played such a fundamental role in formulating [291, 292, 294, 295, 296].[6] Instead, the major focus of this book is on the latter two bases, the regulative or normative status of the norms as the embodiment of ideal standards of rationality and the epistemic status of the norms.

With regard to the regulative status of the current norms of science, the contention of this book is that, as reasonable as those norms appear, they are not synonomous with the essence of scientific rationality; there are aspects of scientific rationality which the current norms seriously fail to capture. And further, these aspects of rationality can not be accounted for entirely in terms of the competition or ambivalence among various elements of the current set of norms, as the sociologists of science have attempted to argue in order to preserve the norms [423]. In other words, contrary to some arguments[7] in the writings of the sociologists of science [423], the current norms are not the only possible norms of science. They are neither unique nor sufficient for the attainment of scientific rationality, although I assert that they are necessary in a sense that is argued for later.

Moreover, this book is particularly critical of the intellectual arguments — the evidential bases — that have been used to infer the traditional norms and to perpetuate their subsequent consideration. It can be shown that many of the reasons that have been used to argue for the necessity and existence of the current norms are circular, if not outright tautologies. For example, consider that deviation from the norms is explained in terms of competition or ambivalence among the norms themselves. Consider, too, that deviation from the norms has been taken as confirmation of the norms [297, pp. 607—608]. If every deviation is either the result of ambivalence or competition among norms, or if every deviation actually reaffirms the norms, then what do the norms explain? Are they not tautologically true? Are they not only invoked to explain and also to justify themselves?

I am not objecting to the existence and use of tautologies per se, for it can be argued that the underlying fundamental principles of every science become tautological at some point [72, 77, 79, 81]. The lack of reflectiveness about the use and perpetuation of tautologies is objectionable, not the tautologies. Without this reflectiveness, one is in a weak position to assess when one's tautologies are no longer fruitful.

Finally, another line of argumentation has been used to support

14

these norms — the distinction between the institutional and motivational levels of analysis. Presumably the current norms apply only to the social, institutional level of science taken as a whole; they are not to be taken as psychological properties or correlates of individual scientists. For example, consider the following passages from Merton:

> Science, as is the case with the professions in general, includes disinterestedness as a basic institutional element. Disinterestedness is not to be equated with altruism nor interested action with egoism. Such equivalences confuse institutional and motivational levels of analysis. A passion for knowledge, idle curiosity, altruistic concern with the benefit to humanity, and a host of other special motives have been attributed to the scientist. The quest for distinctive motives appears to have been misdirected. *It is rather a distinctive pattern of institutional control of a wide range of motives which characterizes the behavior of scientists.* For once the institution enjoins disinterested activity, it is to the interest of scientists to conform on pain of sanctions and, in so far as the norm has been internalized, on pain of psychological conflict [297, pp. 612—613].

Taken literally, Merton's prohibition is absurd, for Merton himself does not, and indeed could not, keep the two levels completely distinct (and therefore nonconfused), no matter how much he desires to do so. Indeed, it is a trivial matter to point out the psychology that Merton has had to presuppose in order to make the argument that the norms do not "largely" rest on psychological considerations. For example, the institution of science may enjoin "disinterested activity," but it can be shown that not every psychology is receptive to or can conform to the norm, particularly "on pain of psychological conflict" [128, 203—205, 285, 456].

The stricture to keep the two levels distinct conveniently ignores the fact that the norms were built up, in the first place, by inductively generalizing from the cases of individual scientists [297], from the highly select writings of the rare, great scientists who were psychologically motivated enough to write glowingly of science. That the great scientists characterized science in the highly idealized ways that they did is, in the very least, mute testimony to their psychology. *Merton would thus enjoin us to refrain from confusing what he has already confused or presupposed implicitly.* Apparently it is all right if we (a) adopt the implicit, idealized attitudes (psychology) of the great scientists, (b) elevate that psychology to institutional norms, (c) posit a sharp distinction between the institutional and motivational levels of analysis, and thus, (d) make the institutional level immune to challenge by

15

the individual, motivational level. In other words, apparently it is all right to base the institutional norms of science on the idealized attitudes of the great scientists, but it is not all right to critically test these idealized norms against the messy behavior and complicated attitudes of real scientists. But if we are justified in using idealized attitudes to build up these norms, why can we not use other attitudinal evidence, if not to tear them down, at least to critically reexamine them? If we cannot, then we are back to the criticisms of the previous paragraphs; we are guilty of fostering tautological statements. More specifically, we foster a bad metaphysics that is immune to all testing or challenge by one of the strongest sources that could conceivably challenge it — the actual attitudes and behavior of real scientists [132—134].

Other criticisms[8] can be lodged against the reasonableness of these norms and the arguments that have been used to derive and to uphold them [33, 297, 423]. However, one of the central issues of this book concerns the bearing of countervailing empirical evidence on normative claims in general. That is, it is questionable whether normative claims cannot be tested by empirical evidence of any sort. Normative claims may be immune to the kind of empirical evidence that merely bears on whether the norms are realized in practice or not, but they are not necessarily immune to the kind of empirical evidence that consists of the serious and interesting theoretical arguments of a group of contemporary scientists for the reasonableness of scientific behavior that runs sharply counter to the current norms of science. As I demonstrate later, one of the most interesting outcomes of this study has proven to be a strong case that can be made for the existence of a set of counter-norms. Each of these counter-norms can be shown to stand in opposition to the current norms of science.

More than anyone else, Merton has seen the need for counter-norms. In a number of essays [291, 292, 294, 298], Merton has repeatedly focused on the notion of sociological ambivalence as a major, if not a dominant, characteristic of social institutions; as such, this ambivalence is central to the analysis and understanding of social institutions:

> One characteristic of social institutions is that they tend to be patterned in terms of conflicting pairs of norms...
> Did my time and your patience allow, it would be possible to consider, first, how potentially contradictory norms develop in every social institution; next, how in the institution of science, conflicting norms generate marked ambivalence in the lives of scientists; and finally, how

this ambivalence affects the actual, as distinct from the supposed, relations between men of science [291, pp. 78, 80].

At issue is how we conceive the structure of social roles. From the perspective of sociological ambivalence, we see a social role as a dynamic organization of norms and counter-norms, not as a combination of dominant attributes (such as affective neutrality or functional specificity). We propose that *the major norms and the minor counter-norms* alternatively govern role-behavior to produce ambivalence [298, p. 103; emphasis added].

And finally:

From the standpoint of sociological ambivalence..., the structure of [for example] the physician's role [consists] of a *dynamic alternation of norms and counter-norms* [emphasis added]. These norms call for potentially contradictory attitudes and behaviors. For the social definitions of this role [the physician's], as of social roles generally, in terms of dominant attributes alone would not be flexible enough to provide for the endlessly varying contingencies of social relations. Behavior oriented wholly to the dominant norms would defeat the functional objectives of the role. Instead, *role behavior is alternatively oriented to dominant norms and to subsidiary counter-norms in the role* [emphasis added]. This alternation of subroles *evolves* as a social device for helping men in designated statuses to cope with the contingencies they face in trying to fulfill their functions. This is lost to view when social roles are analyzed only in terms of their major attributes [288, p. 104].

If you take Merton's ideas regarding sociological ambivalence seriously, as I do, then it is misleading, if not outright false, to think of any institution — including science — as governed by a single set of norms. Instead, one is directed to look for competing sets of norms. As a result, explicating these opposing sets of norms and understanding their relationship to one another becomes one of the major tasks of the sociology of science.

For the preceding and other reasons, it will be argued that at least two sets of norms are necessary for the rationality and growth of science, and that the ambivalence and competition between these sets of norms are necessary to account for the tension inherent in the structure of scientific knowledge and in the institution of science. For example, the tension manifests itself in a competition and ambivalence between the norm of disinterestedness on the one side and the norm of interestedness on the other. If there are good reasons why scientists should be disinterested, then there are also good reasons as to why they should be interested — for example, to develop strong commitments to their favorite hypotheses and theories.

The words of physicist A.M. Taylor provide a fitting epilogue to the material of this section:

> [According to Mach] above all, *scientists must be prepared immediately to drop a theory the moment an observation turns up to conflict with it. Scientists must have an absolute respect for observations; they must hold scientific theories in judicial detachment. Scientists must be passionless observers, unbiased by emotion, intellectually cold.*
>
> The facts are otherwise. The history of science shows us, again and again, great discoveries made by passionate adherence to ideas forged in the white heat of imagination. It shows us slow construction, brick by patient brick, of a scientific edifice, often in complete disregard of apparently conflicting evidence. It shows us bold imaginative leaps made in the dark, in unjustified anticipation of success, only later to receive astonishing experimental confirmation.
>
> *The three attributes of commitment, imagination, and tenacity seem to be the distinguishing marks of greatness in a scientist.* A scientist *must be* as utterly committed in the pursuit of truth as the most dedicated of mystics; he must be as pertinacious in his struggle to advance into uncharted country as the most indomitable of pioneers; his imagination must be as vivid and ingenious as a poet's or a painter's. Like other men, for success he needs ability and some luck; his imagination may be sterile if he has not a flair for asking the right questions, questions to which nature's reply is intelligible and significant [433, pp. 4—5; emphasis added].

Logic of Research Versus Social Psychology of Research

Abraham Kaplan [219, p.12] wrote: "The difficulty is that often in conventional logic much of what actually goes on in science is dismissed as belonging to psychology or sociology, until it has been transformed to suit an antecedently chosen reconstruction." The bearing of empirical evidence on the status of normative claims is only a small part of a much larger issue. The larger issue is that of the bearing of the claims, patterns of reasoning, and evidence of one discipline or field on that of another. A major contention of this book is that one of the reasons that the various images that we have of science are so limited is that they are the product of extreme disciplinary thinking. For example, one of the reasons the current norms of science are so limited in scope (rationality) is that they overly reflect the thinking of one discipline, sociology. The current norms have been given over almost exclusively to sociology as its problem to work on, when in fact the

18

definition of the norms of science can be construed as a problem for every discipline that has an interest in explicating the nature of science [16, 70—72, 79]. Viewed from this perspective, the fault does not lie exclusively with sociology. The various disciplines are as much at fault for having turned their backs on the formulation of the norms of science as a problem for their science.

The basic problem is a philosophical one in the sense that the way one views the relationship between the various disciplines is itself expressive of a philosophical position [72, 79, 408]. Consider the matter in the following way: If it can be argued that a particular scientific theory is a policy — in effect a directive — with respect to how one ought to view and to study nature [72], then I would argue that a particular theory about the nature of science is also a policy, in this case, a policy with respect to how one ought to study and to characterize science. The physicist J.J. Thompson put it as follows: "From the point of view of the physicist, a theory of matter is a policy rather than a creed; its object is to connect or coordinate apparently diverse phenomena, and above all to suggest, stimulate, and direct experiment" [439, p. 1]. In other words, a particular theory about the nature of science is obviously not a factually real description of science as it is. Rather, it directs those who hold the theory to seek a particular kind of explanation of science. But if so, then a particular theory about science is at the same time a partial state description of its adherents. As much as it tells us something about the nature of science (which it does), it also tells us something about the nature of the proponents of the theory. It is expressive of their commitments to a particular point of view. This is certainly one way that the recent dispute between the logic of research and the social psychology of research [246, 348] can be viewed.

The logic of research, or the logic of science as it is more commonly called, is concerned with utilizing the powerful methods of symbolic logic to construct a formal systemic account of science. The overwhelming emphasis of this approach is on the pure logical form of scientific theories and statements rather than on their detailed empirical content. Indeed it is an implicit tenet of this approach that underneath all of the confusion of everyday scientific practice there is a stable substructure that can be culled out and explicated by the powerful techniques of logic. In short, the emphasis of this approach is on a formal, axiomatic, deductive account of scientific theories. This approach has placed a strong emphasis on a crucial distinction, the distinction between the context of discovery and the context of justification [358, 359]. Whereas the context of justification, or that phase of scientific activity concerned with the validation or testing of scientific ideas, is supposed to be susceptible to logical explication, the context or process of discovery is supposed not to be. Accordingly, the concern of the logician is to be focused almost exclusively on the justification or testing of ideas; discovery is relegated to second place and shunted off for the psychologists to study. Even more distressing, it has been asserted that the logician need not concern himself at all with the aspects of discovery. How a scientist has actually discovered his ideas is supposedly irrelevant.[9] All that counts is how those ideas are tested.

In contradistinction, the social psychology of research tends to emphasize the inseparability of discovery and justification [246]. Above all, it emphasizes the social and psychological nature of scientific work. It stresses that scientists are not brought up and trained to approach scientific problems within a formal, logical tradition, but that instead they are socialized into the traditions of their respective fields through solving countless characteristic problems or puzzles [247, 248]. The influence of these puzzles is so great that they provide the scientist with the characteristic methods, or heuristic rules, that he will use to solve the professional problems he will later encounter, and they also shape his sense of professional identity. That is, he will be able to identify easily with those who have gone through a similar mode of education, and he will find it difficult to identify with those who have not. Unlike the logicians' account of science, which stresses the abstract-rule nature of science, this account instead stresses the informal social and psychological processes that govern scientific life. It is dedicated to explicating them. In the extreme, it not only denies the utility of logical accounts of science but also their very possibility, let alone their validity.

Whatever the final outcome of the debate between these two currently contending approaches to the study of science, it is my feeling that the present tendency to view the logic of research and the social psychology of research as locked in to an adversary or zero-sum game situation is more indicative of the current state of our philosophy of science than it is of science, logic, and social psychology. Contrary to the belief expressed by some in the philosophical literature, it does not follow that the only way to view the relationship between logic and social psychology is as an adversary one or one of a superior/inferior relationship.[10] I would argue not only that there are other ways to view the relationship, but that even more important, that the failure to take seriously some of these other ways has actually impeded some interesting new lines of research and developments in the philosophy of science. Even worse, it has had the disastrous side effect of excluding before the fact some of the strongest lines of evidence which could most effectively challenge some of the prevailing points of view.

It would be considerably easier to accept the arguments of those who have been most severe in their condemnation of the social psychology or research and of its potential contributions to the philosophy of science were it not for the following: (1) In every case it is possible to show [309] that that position rests on a particular, largely implicit, social psychology — that is, on a number of propositions, regarding the nature of science and scientists, that are sociological and psychological in their import.[11] (2) For the most part this implicit, homespun social psychology is inferior to the professional products already available on the market (that is, in the professional literature). And (3) most of the critics of social psychology have exhibited a gross misunderstanding of the aims of the social psychology of research and a gross ignorance of the facts and methods of social psychology.

It is little wonder, then, that the view of social psychology that has been promulgated most often has tended to be the worst characterization of social psychology possible. For example, it is a gross misunderstanding of the aims of social psychology, and it is basically demeaning of it, to say that the potential contributions of social psychology are limited to uncovering mere matters of empirical fact or that it is perpetually relegated to studying only those aspects of science having to do with the messy, illogical aspects of discovery. It is also a misunderstanding to perceive social psychology as having little if anything to contribute to the study of the epistemology of testing scientific ideas. Given these perceptions of the role of social psychology, it is not surprising that any appeal to the social psychology of research has tended to be met with charges of sociologism or psychologism [318, 347—349].

One of the major purposes of this book is to demonstrate what the social psychology of research uniquely has to contribute to the philosophy of science. It is the contention of this book that there are some contributions that are beyond the logic of science, at least as it is currently constituted, to make. This book thus enters centrally into the recent dispute between Thomas Kuhn and Karl

Popper [246, 348] regarding the relative merits of the logic of research and the social psychology of research.[12] In general this book is supportive of Kuhn's position, critical of Popper's. However, it will be shown that Kuhn's position is in need of some fundamental revisions if it is to meet some of the criticisms that have been legitimately lodged against it [396, 397]. Some of the same revisions will also be shown to apply to Paul Feyerabend's positions. The entire study mounts a severe critique of the orthodox [129, 180] or received [426] view of scientific theories. Thus, it is very much in sympathy with Stephen Toulmin's latest efforts [446]. In particular, it will be argued that the orthodox or received view of scientific theories is found wanting in several respects: (1) scientific observations are not theory-free; (2) scientific observations are not directly observable; (3) observations are no less problematic than theories.

C. West Churchman has consistently articulated a philosophy of science [71, 72, 77, 79, 81] wherein the logic of research and the social psychology of research are taken as vital but partial components of the program. In other words, neither the logic nor the social psychology of research form the whole, or *the* core of the program. Indeed, no single science forms the core or becomes *the* single basis for explicating science.[13] In a word, Churchman's philosophy of science is characterized by the fact that it is fundamentally "anti-reductionistic." All of the sciences now known are conceived of as indispensable in the sense that there is an aspect of the process of science that each of the various sciences is uniquely suited to uncovering and to studying. Viewed from this perspective, the relationship between the logic and the social psychology of research is a complimentary one, not a hostile adversary relationship, although it is vitally necessary that each of these components be free to criticize the other.

In terms of this view of science, the Reichenbach distinction [358, 359] between the context of discovery and the context of justification is not only naive — it is actually harmful to studies into the nature of science. The distinction is regarded as naive, because a Churchmanian view of science does not recognize that scientific activity can be fundamentally partitioned into discovery versus testing or justification phases; it does not recognize that there is a sharp cleavage between the acts of discovery and of testing, and as a result, it does not see where the process of discovery can be explicated *even in theory*, independently of the process of testing, and vice versa. As a consequence, a Churchmanian view also regards the distinction as harmful because it promotes a separatist, piecemeal view of science in theory and in practice by, for example, discouraging the kind of broad-based, interdisciplinary studies which could uncover the necessary evidence that could effectively challenge the basis for the distinction. This is perhaps one of the most distressing things about strict disciplinary thinking. The disciplines erect such formidable barriers to the arguments, methods, and data of other disciplines that they insulate themselves from the strongest possible challenges that could be mounted against their most basic concepts.[14]

The study on which this book is based is an example of a Churchmanian program in the philosophy of science. It is founded on the premises (1) that the philosophy of science has important contributions to make to social science theories and studies of science, and (2) that social science studies, theories, and data have important contributions to make to the construction of philosophical theories of science. The purpose of this book is to demonstrate the vitality and validity of a philosophically based social science study of science and also to demonstrate the vitality and validity of a social science based philosophical study of science.

Personal Commitment

To recapitulate, one of the major contentions of this book is that, far more than we realize, we influence the objects we think we study so dispassionately. This is true no less of science than it is of any other field of human inquiry. It is one of the major purposes of this book to illuminate the bounds and nature of this influence. Thus, instead of pretending that this influence does not exist, and so closing my eyes to it, I have actively looked for it. As a direct result, I have not suppressed my own emotions or my own sense of commitment. I cannot in good conscience pretend that what is present in others, and what I have deliberately studied in others — in particular the scientists of this study — is absent in me. I have deliberately chosen not to suppress my own intellectual commitments, because for too long one of the myths we have lived with is that science is a passionless enterprise performed by passionless men, and that it *has* to be if it is to be objective. What this myth ignores is that many of the great scientific achievements of the past have been the result of passionate, if not outright biased, inquiries. I wish to show, that science is no less objective

23

because of this passion. Indeed, there are serious reasons for contending that science is more, not less, objective *precisely because of* (and not in spite of) the presence of great passions. Finally, I have chosen not to suppress my own passions because it was a kind of passion, or emotional commitment, that first generated and then sustained this study. I seriously hope this document is more, not less, objective because I have chosen to disclose my commitments rather than hide them behind a veil of jargon and seemingly impersonal methodological safeguards and rules. I have thus chosen to expose my commitments to criticism and evaluation in order to encourage others to do the same. If more of us were to do so — that is, if the exposure of one's commitments were to become a standard methodological practice — who knows how much more we would know about how we actually do science? In any event, for me to adopt the style of the very myth this book is challenging would be hypocritical. As Churchman has put it:

> One of the most absurd myths of the social sciences is the "objectivity" that is alleged to occur in the relation between the scientist-as-observer and the people he observes. He really thinks he can stand apart and objectively observe how people behave, what their attitudes are, how they think, how they decide. If his intent were to be the clown rather than the objective scientist, we could appreciate him more, because in some ways his own behavior and the manner in which he describes the behavior of others is hilarious.
>
> Instead of the silly and empty claim that an observation is objective if it resides in the brain of an unbiased observer, one should say that an observation is objective if it is the creation of many different points of view. What people are really like is what people with the strongest of inquiring motivations will perceive themselves to be like. The "verification" of a scientific finding resides in the creative spirit of human inquiry carried to its maximum potential [70, pp. 86—87].

If the myth of detachment can be shown not to hold within the physical sciences, it surely can be shown to hold even less within the social sciences.

This book is not a brief for or against science. I am examining some of the crucial underpinnings of science as objectively as I can. I consider myself too much of a scientist to want to see the scientific spirit and temperament disappear from this world of ours. But I also consider myself too much of a reflective spirit not to want to see many of science's crucial assumptions drastically examined and changed. If we cannot examine science as a subject with the same sense of passion that we examine other things, and

24

if science basically requires a set of crucial, unexamined myths for its continued existence, then I seriously question whether it is worth preserving. I am interested in neither the mindless destruction nor the mindless preservation of science. Both are attitudes that should be repugnant to any intellectual.

This book is a critical examination of those who have written about what science is and is not. It is not an attack upon the scientists who so graciously consented to and supported this study.

Tentative Moral and Recapitulation

Why go to the moon? Why study the moon scientists? No two questions could be more fundamental. And they are more closely and intimately related than one might suspect.

So much has been written and spoken regarding the first question that it may seem that the reasons for our having gone to the moon are so perfectly obvious that they require no further elaboration or justification. The reasons are so well known that it seems only necessary to list them briefly. Thus, it has been offered that it was in our "national destiny" to go to the moon; that we went "because it was there"; that our national purpose and international interests required that we go; that the moon was "the new sea of exploration and it was our duty to sail her"; that we were in a race to beat the Russians; and, finally, that the moon would yield unlimited scientific and technological possibilities and knowledge. Whereas the answer to the first question seems so perfectly obvious, the second question seems so esoteric as to be of interest only to the extreme specialist. Yet, as natural as these responses are, they are dangerously deceptive.

The reasons that were so obvious (if indeed they ever really were) for launching our trip in the early 1960s no longer suffice for the 1970s. The country's social problems have become too serious and too pressing, and we have undergone one too many political tragedies to be satisfied any longer by the old and obvious reasons. If we are to continue to justify having gone to the moon, we will have to find not only new reasons but deeper and more reflectively satisfying reasons. After all, what *could* justify such enormous expenditures, and not just expenditures of money alone but, more important, the expenditure of vital psychic resources. In the end the greatest expenditure may reside in the fact that the

25

false belief in technology was perpetuated and kept alive by yet another generation of eager believers — that we once again trained another generation of technologists and proved to them that, with enough money, technical problems could be solved, whereas social problems could not be solved with that same amount of money. This time the technologist's self-fulfilling prophecy would be enacted on the cosmic level.

If the obvious reasons no longer satisfy, we are compelled to look elsewhere. Could it be that the only answer that could ever begin to satisfy the first question was in some mysterious way linked to the second? Could it be that the real reason that we went to the moon, although it was completely unknown to us at the time, was to illuminate a side of ourselves that was as hidden as the far side of the moon?

Chapter Two

Methodology:
Prelude to Apollo 11 Interviews

This chapter and the next deal with the material that was gathered from the first round of interviews. This chapter deals mainly with the methodology that was used to gather the interview material, the next with the presentation and analysis of that material. This chapter (1) indicates how and why the issues that were used for exploration were chosen and generated; (2) gives a preliminary indication of the specific sample of scientists — how they were chosen and their background characteristics; and (3) indicates how the interviews were conducted — the methodology underlying the results in this and the next chapter. The next chapter (1) discusses the interview results and shows how they relate to the issues that have been raised in the introductory chapter, and (2) gives an indication as to how the later interview rounds were built on the results of the first round.

The chapters dealing with the results of the interviews are organized around each interview round instead of by topic area. Thus, Chapter Three presents the results of the first round of interviews; Chapter Four, the second round, and so on. This manner of presentation is not only well suited for presenting the results of the interviews but is also helpful for indicating how to conduct studies of this kind. This is the sense, referred to in Chapter One, in which this is not only a book about the nature of science but also a book about how to study science.

Logistics of Interview Rounds

The first round of interviews was begun on November 25, 1969, and was completed on May 22, 1970. Thus, these interviews began almost four months after the landing of Apollo 11 (July 20, 1969) and took six months to complete. Forty-two scientists distributed throughout the U.S. were interviewed during this period. All of the interviews were tape-recorded for later transcription and analysis. The shortest interview was 45 minutes. The longest was three and one-quarter hours. The average length was two and one-quarter hours. The first round of Apollo 11 interviews consumed approximately 95 hours of recording time and involved approximately fourteen hundred pages of transcript.

The use of a tape recorder was decided on for a number of reasons. It lent a degree of permanence to the interviews. Many of the interviews are with outstanding and important scientists. If only for historical reasons, the recordings are worth preserving. They certainly are one way of recording the history of the Apollo missions, at least as far as that history is reflected through the comments of some of its most important participants. The tapes constitute a set of permanent historical documents which possess some advantages that other methods of recording do not. With tape recordings, one can go back and reanalyze the data in ways not possible with written methods of recording. To cite just the most obvious example, tape recordings capture and preserve the detailed vocal inflections of the respondents. It is pertinent to note in this regard that preliminary arrangements have been made with the Center for the History of Physics, American Institute of Physics, New York City, for depositing the interviews with the scientists there, so that other scholars will have access to the tapes and the written questionnaires.

Figure 1 shows the time periods over which each of the interview rounds (I through IV) was conducted. Also shown are the times of the various Apollo missions (11 through 17). The extreme right column shows the Apollo missions which were covered by each interview round. Thus, for example, even though prior to its start round I was preceded by both Apollo 11 and Apollo 12, round I effectively contains the reactions of the scientists to the Apollo 11 data exclusively. This was due to the close spacing of the missions.

The scientists felt that they could barely begin to analyze the rocks from one mission, let alone assimilate the full range of analyses performed, before the rocks and analyses of another mission were bearing down on them. A number of the scientists remarked sarcastically that this was one indication of how low a priority science occupied in the planning and conduct of the missions. The missions, so they contended, were not spaced for the con-

28

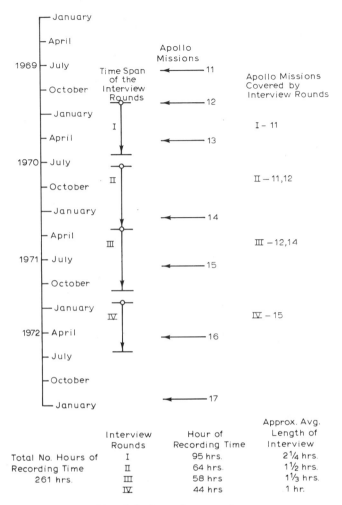

	Interview Rounds	Hour of Recording Time	Approx. Avg. Length of Interview
Total No. Hours of Recording Time 261 hrs.	I	95 hrs.	2¼ hrs.
	II	64 hrs.	1½ hrs.
	III	58 hrs	1⅓ hrs.
	IV	44 hrs	1 hr.

Figure 1. Timetable of the interview rounds.

venience of the scientists but rather to maintain the interest of the public.

During the short time span between Apollo 11 and Apollo 12, the scientists were so busy preparing their analyses of the Apollo 11 rocks for presentation at the first Lunar Science Conference, held in Houston, January 5—8, 1970, that they could not turn their attention to the Apollo 12 rocks. Although the specific situation varied from round to round, the general result was to delay the effect of each Apollo mission on the attitudes of the respondents by at least one interview round. This, plus such factors as the order in which the respondents were interviewed, served to make the influence of the Apollo missions uniform over any one interview round. Although this may have not been desirable from the standpoint of the participants, it was most desirable from the standpoint of the study.

The average length of interview, as shown in Figure 1 is of interest here. Each round called for the same amount of information. Thus, the decrease in time is not due to decreasing amounts of information asked for. Round IV, in fact, probably asked for the most information from the respondents but took the least time to go through. The big decrease between rounds I and II is due to the fact that from round II onward paper and pencil questionnaires and instruments administered in person were used to obtain the responses of the sample. The first big decrease, in other words, is due to the use of more structured ways of collecting information than were used in the relatively unstructured interviews of round I. The succeeding decreases can be explained by the generally improved efficiency of both the interviewer and the respondents in proceeding through the interviews. In addition, by the fourth round, I was doing everything in my power to keep the drain on the scientists' time to a minimum. By that time it was felt that any interview requiring more than an hour of the scientists' time would be difficult to obtain.

General Characteristics of Sample

None of the scientists who participated in this study will be mentioned by name. Anything that would serve to identify a specific respondent must be suppressed, particularly anything that would serve to identify the comments of one individual about another. The majority of the comments were communicated to this author with the strict understanding that the identity of the respondent would not be revealed. However, it is important to note that only in a very few cases did a respondent place any restriction on the use of his comments. The prohibition was generally on the identification of respondents with their comments, not on the use of the comments. However, this limits description of many of the sample characteristics normally discussed in a study of this kind.

Table 1 gives the breakdown by interview rounds and by institu-

Table 1

Institutional Affiliations of the Sample

Institutional affiliation	Interview round							
	I		II		III		IV	
	Per-cent	Num-ber	Per-cent	Num-ber	Per-cent	Num-ber	Per-cent	Num-ber
University or university affiliated research laboratory	62.0[a]	26	60.0	27	59.1	26	60.5	26
NASA installations	21.4	9	24.5	11	25.0	11	23.3	10
USGS plus related govt. agencies	7.1	3	6.7	3	6.8	3	7.0	3
Govt. research laboratories and institutions	2.5	1	2.2	1	2.3	1	2.2	1
Private industry	7.1	3	6.7	3	6.8	3	7.0	3
Total [b]	100.0	42	100.0	45	100.0	44	100.0	43

[a] 26/42 = 62.0%
[b] 37 scientists were interviewed 4 times, across *all four* rounds.
43 scientists were interviewed 3 times, across the *last* three rounds

tional affiliation of the numbers of scientists interviewed. Thus, in the first round, 26 university affiliated scientists were interviewed, 9 NASA affiliated scientists, 3 United States Geological Survey (USGS) affiliated scientists, 1 scientist in a government related institution, and 3 scientists who were located in private industry. [1] Of the 26 original university scientists in the first round, 22 were interviewed across all four rounds. Four of the original 26 dropped out in the second round, and 5 new scientists were added. Of the 5 new scientists added in the second round, one dropped out in the third round. Two additional NASA affiliated scientists were added in the second round; these were then subsequently interviewed for the two remaining rounds. Finally, one NASA scientist dropped out in the last round. The table indicates that 37 scientists were

31

Table 2

Proportion of the Sample Involved in the Direct Analysis of the Apollo 11 Lunar Materials

Institutional affiliation	Form of involvement in the lunar program				
	PIs	Co-Is	Sample access	No-Is	Total numbers
University or university affiliated research laboratory	[a]57.8% 15 [b]83.3%	7.6% 2 50.0%	15.4% 4 40.0%	19.2% 5 50.0%	61.9% 26
NASA installations	[c]0% 0 [d]0%	22.2% 2 50.0%	44.5% 4 40.0%	33.3% 3 30.0%	21.5% 9
USGS plus related govt. agencies	[e]33.3% 1 [f]5.6%	0% 0 0%	33.3% 1 10.0%	33.3% 1 10.0%	7.1% 3
Govt. research laboratories, institutions	[g]100.0% 1 [h]5.6%	0% 0 0%	0% 0 0%	0% 0 0%	2.4% 1
Private industry	[i]33.3% 1 [j]5.6%	0% 0 0%	33.3% 1 10.0%	33.3% 1 10.0%	7.1% 3
Total numbers	18 [k]42.9%	4 9.5%	10 23.8%	10 23.8%	100.0% 42 100.0%

[a]15/26 = 57.8%; [b]15/18 = 83.3%; [c]0/9 = 0%; [d]0/18 = 0%; [e]1/3 = 33.3%; [f]1/18 = 5.6%; [g]1/1 = 100.0%; [h]1/18 = 5.6%; [i]1/3 = 33.3%; [j]1/18 = 5.6%; [k]18/42 = 42.9%

Note: The figures and percentages are reported only for the scientists in interview round I (Apollo 11) because the numbers are virtually the same for the remaining rounds.

interviewed across all four rounds and that 43 scientists were repeatedly interviewed for at least three rounds, that is, across the remaining three rounds.

Table 2 gives the breakdown by institutional affiliation of the numbers of scientists interviewed in round I (the time period of the Apollo 11 mission) who were: (1) principal investigators (PIs) or (2) co-investigators (Co-Is), (3) those who were neither PIs or Co-Is but who had access to or contact with the lunar samples in some form or another (Sample Access), and finally, (4) those scientists who had no contact at all with the lunar samples (No-Is). The term PI is the official designation that NASA (in conformity with the practice of most granting agencies) used to denote the principal researcher (proposer) of a project or experiment. Every experiment, whether or not it had a co-investigator (Co-I), had a principal investigator (PI). The term Sample Access is my own. It denotes all those scientists who were neither PIs nor Co-Is but who nevertheless had legitimate access to the lunar materials as being members of a PI's research team. These scientists made a direct contribution to the experimental study of the lunar samples. This does not apply to the remaining category of scientists. Although many of them were indirectly involved, the No-I scientists (my term) were scientists who, for the most part, were not involved with the analyses of lunar materials. They were interviewed because they had played an important historical role in our understanding of the moon, because they had a unique perspective or insight into the understanding of the lunar missions, of their fellow scientists, or because they had been highly recommended by their fellow scientists for inclusion in the sample.

Table 1 shows that the highest percentage of scientists interviewed in every round were university-based scientists. Table 2 shows that the highest percentage of PIs in round I (Apollo 11) were again university scientists. Because Table 2 is representative, this generally holds true of all the interview rounds. To appreciate the significance of these and the remaining percentages, one must understand how the sample was formed.

The high percentage of university scientists in the sample of scientists studied is not due entirely to the fact that the highest percentage in the total population of scientists selected by NASA to be PIs and Co-Is for Apollo 11 were university scientists.[2] The major contributing factor to the formation of the sample, irrespective of the various percentages of scientists in the various categories, is that *the sample was essentially generated by means of the sample itself.* The sample of scientists that formed the object of this study was not a random sample from the beginning, nor was it ever intended to be. The sample was generated by starting with a very few key scientists who were sympathetic to the study and who were willing to help get it started by lending their name as an introduction to the rest of the sample. Starting with these few key scientists, two questions were asked in turn of every individual who was subsequently interviewed: "Which scientists would you recommend that I ask these same (interview) questions of next?" and "For what reasons do you recommend that I see these persons?"

The reasons for adopting this method of forming the sample were as follows: (1) Given the fact that a number of the scientists in the system had been literally besieged by reporters asking for interviews, and that this was an inordinately busy time in their professional lives, I wanted a way of getting into the system that would induce the respondents to give me their precious time and their serious consideration. (2) Given the fact that a good many of the interview questions were sensitive, I wanted a way of approaching the respondents that indicated to them that it was all right for them to talk about these issues with me. In effect, I wanted them to know that I had been previously certified by other members of their community as someone to whom they could open up. (3) No study of the Apollo missions could afford to ignore a number of important scientists who were in the system. Whether a random sample selected these persons or not, some individuals could just not be overlooked. If it was a choice between a random sample and these individuals, then the random sample had to give way. (4) Given that one of the things I was interested in studying was the social organization of the system, it would have been foolish not to have used the selection of the sample to inform me in some way about this important facet of the system. Any study must be carefully designed to serve the basic purposes of the study [16, 72, 80]. The selection of a sample is one of the key variables that an investigator can manipulate in giving him information about his sample. Because the time these men had was severely limited and thus I only had so much interview time with them, I wanted *every* question that I might ask of them to serve not only the general purposes of the study but as many detailed purposes as possible.[3] Thus, asking the respondents who in their opinion I should interview next not only served to form the sample but also gave me vital information about the sample just by the way it was formed. (5) As an aside but nevertheless important consideration, forming the sample in this manner tends to offset some of the objections [252] that have been raised against studying the average or, in Kuhn's terms, the "normal" scientist. The objection, or so goes the argument, is that a sample composed entirely of average scientists is an inherently poor sample on which to form conclusions about the general nature of science. It can be an even worse basis on which to form

34

conclusions regarding the ideal practice of science. Why, goes the argument, base ideas for the superior or improved practice of science on the behavior of average or mediocre scientists. Selecting a sample in the manner described counters this objection, because such a sample will inherently tend to contain more than its share of the elite scientists within the system under study. That is, one would expect the elites to be recommended more often than not in such a procedure.

If the general population of Apollo 11 scientists represents an elite body of scientists by containing more than its fair share of outstanding scientists, then in more than one sense the sample is best described as an elite of elites. By almost any standard, the sample contains some of the most distinguished and prestigious of scientists. It contains some of the most distinguished members of the geologic profession and certainly some of the most outstanding scientific analysts of the Apollo missions. Six of the scientists are members of the National Academy of Scientists; 38 out of the 42 scientists, or 90.5 percent, have their Ph.Ds. Thirteen hold major editorships on key scientific journals in the field.

This does not mean that every scientist in the sample was distinguished or prestigious. In fact it is both interesting and significant that a good many of the scientists based their recommendations as to which other scientists I should see on three main criteria. First, most of the scientists indicated by their comments that they assumed I would see or already had seen other scientists of obvious importance or eminence. Second, the vast majority of the scientists recognized that they each represented a particular point of view, and it is to their credit that most of them recommended that I see scientists with opposite views so that the study would capture as many points of view as possible. Third, and this is closely related to the second criteria, in order for the study to be representative in another sense, I believed the sample should contain a number of scientists who were judged to be typical or representative of the average scientist in the Apollo program. Many of the scientists expressed this point with comments such as: "You should see some of the average stiffs, not just the stars," or, "If you see only those guys, you've got a sample of all chiefs and no Indians."

The mean age of the sample as of July 1969 (the time of Apollo 11) was 47.0 years with a standard deviation of 9.3 years. This also is indicative of the way the sample was constructed, and it reflects on the fact that we are dealing with men who have securely established themselves within their profession. Another way to put this is to note that there were very few referrals to young scientists. There were fewer than 6 scientists under the age of 40 in the first round of interviews and only 3 scientists under the age of 35 were interviewed throughout the entire study. In this regard the sample differs quite sharply from the general population of Apollo scientists. In their excellent and eminently readable summary of the scientific results of the Apollo missions, Levinson and Taylor [261] note that in the general population "a surprisingly large number of the scientists are in their 30's; only a small percentage are over 50 years" [261, p. 2]. Thus, the sample almost exactly reverses the trend that exists in the general population. Again only the ages of the scientists interviewed in round I (the Apollo 11 mission) are reported, because there are no major differences in the sample characteristics in later rounds.

Thirty-six of the scientists were classified as American. This means they were not only American born but that they had posts at American universities. The remaining six scientists who were classified as foreign were either foreign born but working in this country or foreign born and working in the country of their birth. The foreign category accounted for the highest percentage of additions to and drop-outs from the sample. These scientists were the hardest to interview on a systematic basis because of the difficulty of matching travel schedules.

Table 3 indicates which scientific disciplines were represented in the sample. In general this refers to the discipline in which a man received his highest degree. Nearly all of the scientists are located

Table 3

Scientific Disciplines Represented in the Sample

Disciplines	Scientists Number	%
Geology (general)	7	16.7
Geophysics	4	9.5
Geochemistry	16	38.2
Chemistry	5	11.9
Physics	4	9.5
Astronomy	4	9.5
Engineering	2	4.8
	42	100.0

in academic departments or institutional settings that correspond closely to the academic disciplines in which they received their degrees. The designation "geology (general)" has been used to distinguish those scientists who received their major training and orientation in general field geology and, as a result, consider themselves general field geologists. It appears from the interviews that those who received general geological training have a very strong, if not inordinately strong, identification with the designation "general field geologist," as compared for example, with the designation "geochemist" or "geophysicist." The effect of this strong attachment to disciplinary labels will be made clear in Chapter Three.

Procedural and Methodological Aspects of Interviews

The decision was made early in the conceptualization of the study to conduct the first round in the form of partially structured, partially open-ended or unstructured interviews. The standardized name in the social science literature that comes closest to describing this kind of interview is *focused interview*. The essentials of this kind of interviewing procedure have been well described in the literature by its originators, Robert Merton and his colleagues [299, 300]. The purpose of this kind of interview is to combine the strengths, while hopefully avoiding the weaknesses, of the completely structured interview and the completely unstructured or free-association technique interview [413]. In the focused interview, the interview questions that are asked of each subject are relatively formalized (standardized), while the question-asking procedure is relatively unstructured (open-ended) and thus somewhat variable from subject to subject.

The reasons for adopting the focused interview procedure were varied. Prior to intensive contact with the sample, I did not feel that I had enough awareness and understanding of the range of issues involved to select a particular subset of issues to follow in detail. Without even this knowledge, I certainly did not feel I was ready to construct a structured questionnaire that would permit a quantitative and systematic assessment of the issues, whatever they were. Thus, to some extent the interviews had to be open-ended and free-wheeling to allow me to discover what issues were worth pursuing on future rounds. And in fact, some of the most interesting and important issues were uncovered through the early interviews. Other questions that were formulated prior to any contact with the sample (for example, "How would you rank the various theories for the origin of the moon in terms of their plausibility or probability?") were kept constant throughout the first round — that is, standardized, and thus asked of every respondent.

Because of the highly sensitive nature of many of the questions and issues, it was felt that these special issues would be better raised and pursued within the context of face to face interviews that, as much as possible, took the form of conversations rather than the form of tightly structured interviews. For this reason I avoided the use of a standardized set of written questions to be handed to the respondents for them to fill out. Nor did I hold in front of me a standardized interview sheet on which I recorded the answers of each respondent to each question. Instead, I memorized a general set of questions that I attempted to ask of each respondent. This procedure had both its strengths and its weaknesses. Its major strength was that it allowed me to concentrate a good part of my attention directly on the respondent, instead of sharing it with a printed sheet. The major defect was that, in the course of following up side-issues that each respondent invariably raised, the same order of questioning was not followed for each respondent, and invariably at least one question per subject was somehow omitted.

Because I wanted to study how the respondents' attitudes changed over time as result of the data generated by the Apollo missions, it was important to avoid as much as possible anything in the first interview situation that would arouse the respondents' suspicions or hostility and thus prevent my being able to interview them repeatedly. The purpose of the first round was not only to start the process of data collection but also to establish rapport, so that I could continue to collect data for a longitudinal study of attitude change. It was felt that a structured questionnaire or interview schedule would be premature and would have the adverse effect of turning off those respondents who were particularly resistant to or not used to such instruments. When one's purpose is repeated measurements, particularly in the social sciences, the purpose of the first round may be to sensitize the subjects to the measuring device and the fact of their being measured rather than to make measurements [459].

Finally, I felt that a tightly structured questionnaire or interview schedule would prevent me from getting at one of the most important pieces of information I could collect — how the respondents viewed and characterized the issues in their terms. It is a well-known truism throughout all of science that every approach, instrument, and methodology has its inherent strengths as well as its weaknesses. In the social sciences this is perhaps doubly so. As Webb, and others, have put it:

> Interviews and questionnaires intrude as a foreign element into the social setting they would describe, *they create as well as measure attitudes* [emphasis added], they elicit atypical roles and responses, they are limited to those who are accessible and will cooperate, and the responses obtained are produced in part by dimensions of individual differences irrelevant to the topic at hand [459, p. 1].

The strengths of a well-structured questionnaire or interview are that they permit the relatively easy collection and systematic analysis of large blocks of data. For this and other reasons, well-structured questionnaires and interviews are best suited for the testing of already formulated, well-structured research hypotheses that the researcher brings with him to the sample [413]. Their weaknesses are that they require the respondent to respond to the issues in the terminology and framework of the researcher. This presupposes that the researcher already has a detailed understanding of the issues and that he is able to formulate them in terms that are meaningful to the respondents. The strengths of the open-ended interview are that it permits the respondents to respond in their own terminology and in terms of their own frames of reference. The well-structured questionnaire and interview are best suited for the testing of already formulated hypotheses; the open-ended interview is best suited for hypothesis discovery and formulation. The open-ended interview makes this possible by allowing both the researcher and the respondent to pursue issues at length. It also does this by allowing them to go into side issues that might not arise via any other method. However, open-ended interviews also have their weaknesses. They can be extremely difficult to collate and thus to analyze. If each respondent brings up a completely distinct idea, there are serious problems in organizing the material into any meaningful patterns, and it is difficult to draw any reasonable generalizations from the interview data. The focused interview tries to steer a middle course between these two extremes by capitalizing on whatever information the researcher has at his disposal about the nature of the sample prior to his interviews with them. This prior information is used to formulate the structured aspects of the focused interview — the interview questions. The somewhat free-flowing or unstructured nature of the question-asking procedure is used to uncover and to formulate new questions, issues, and hypotheses about the nature of the sample.

I personally conducted all of the interviews. Each interview was opened with a short introduction designed to explain briefly the nature and purpose of the study. I explained in general terms that I was interested in recording some of the history surrounding this dramatic event in science by interviewing some of the scientists connected with the Apollo missions. I mentioned that very rarely in the history of science have we had the opportunity to interview directly and repeatedly the participants in an unfolding scientific adventure. I also explained that, in order to capture the history of this event, I would like to visit with them a number of times to record how their scientific ideas changed over time in response to the various Apollo landings.

An hypothesis, which was later generally confirmed through the informal comments of the respondents, was that an initial appeal to the history of science would be more successful in securing their interest and cooperation than a frank appeal to the philosophy of science or to psychology or sociology. Most scientists fancy themselves as generally interested in the history of science or at least as amateur historians of their science, and this seemed to be true of this sample.

As part of the introduction to the study, I also gave the respondents a brief description of my own background. I mentioned explicitly that I had an undergraduate degree in engineering physics and a PhD in engineering science. I included this personal information for the purpose of indicating that I was interested in doing a serious study of the issues and that I had the necessary background to do it. I knew that many of the scientists had been interviewed a number of times by the press. Less than complete satisfaction with the coverage by the science reporters came out more than once during the course of the interviews. I wanted to impress upon the scientists that I had both the interest and the background to do a prolonged and intensive study of their attitudes.

During the initial interviews, I tried to get from each scientist as many of his or her past reprints as I could. In this way I could at least generally familiarize myself with the work and interests of each scientist and keep generally abreast of the scientific literature. During the course of the study I subscribed to a number of the professional journals in the field so that I could follow the appearance of the scientists' articles.

Before the actual start of each interview, I explained briefly that I wanted to use a tape-recorder as the means for recording my research data. In no case was the use of the recorder either denied or resisted. Only in three or four cases through the entire course of the study (over all four interview rounds) was I explicitly instructed to turn off the recorder. As one might anticipate, this occurred only at a point in the interviews when a scientist was deeply involved in discussing intense personal reactions to peers and colleagues. In these few cases the respondents mentioned that they did not wish to have their comments go on record, although none of the respondents hesitated to speak aloud to me what they were unwilling to speak into the recorder. My promise to adhere to a strict norm of confidentiality seemed to be accepted at its face value. I emphasized this at the start of each interview, and I usually reiterated it at the point in the interviews when the scientists were asked for their opinions of their colleagues. I emphasized that I was only interested in the substance of their comments. I was not interested in identifying who said what about whom or in identifying who felt what about a particular issue.

The single word that comes closest to describing the style I tried to manifest in conducting the interviews is the word *facilitating*. I tried to convey the distinct impression that I was there for the sole purpose of listening to each man. And I was. I listened to each man as intensely as I could, so that I could assimilate as much of his individuality as possible and especially so that I could understand his point of view. I conceived of my role as an interviewer who was there to get information, not to challenge or to criticize. In the next chapter, when I begin to describe my role as a critic, the situation is very different. The closest I came to challenging was the few times in every interview that I played the devil's advocate for some other point of view. Even then I explicitly mentioned that I was playing the devil's advocate for the express purpose of getting their reaction to that other point of view. Those who are familiar with the general idea of Rogerian nondirective interviewing [413] will recognize some of the facilitating elements that were borrowed from that approach.

All the precautionary procedures and methodological safeguards in the world amount to nothing if one's respondents are basically unwilling to talk. After all was said and done, the basic fact was that the respondents were more than willing to talk; it was as if they not only wanted to talk but that they needed to talk. We shall have more than ample opportunity to examine why in the next chapter.

Initial Interview Questions

In accordance with the methodology of the focused interview, there were two rather broad sets of interview questions that, as much as possible, were used throughout the entire first round with every respondent. The first set consisted of those questions that, prior to any contact with the sample, were formulated as initial probes. The second set consisted of questions that emerged in response to the initial set. By their very nature, the second set of questions could not be asked of every respondent, at least not in the first round, because they only emerged during the first round. In this section, we shall mainly be concerned with the initial set. Here I simply list these questions and comment briefly on their nature and on what they were intended to accomplish. I discuss the second set of questions in the next and later chapters, when I discuss the responses to the initial set and hence at the points where the second set emerged.

Before I list the initial questions, some further comments are required. Both the interview questions that were put to the scientists and their responses to these questions can be organized around three main categories. The first category (questions 1—8 below) deals with those general questions and issues that pertain to *the nature of the moon as a physical object*. These have to do with the nature of the moon's physical properties and with possible mechanisms for its origin. The main question here is how the various scientists perceived the moon's properties just prior to or just immediately after Apollo 11 and how they lined up into various camps — for example, how many scientists favored a hot moon versus a cold moon and for which reasons. The second category deals with questions and issues related to *the nature of science*, both as an institution and as a method for achieving knowledge. Here the question is how the scientists felt about such abstract notions as scientific detachment, emotional neutrality, and objectivity. The third category of questions deals with issues that are related to *scientists as persons*. Here the question is how the scientists felt about one another in personal terms as peers and as colleagues. Who among them were regarded as the outstanding theorists of lunar origin. Who were regarded as the most likely to hold on to their favored theories or pet hypotheses in the face of overwhelming negative evidence. (Questions 9—15 below pertain to the nature both of science and of scientists as persons.) The questions are listed in the general order in which they were asked of each scientist.

1. Before Apollo 11, which theory regarding the origin of the moon did you regard as most plausible? Least plausible? Could you rank/list the theories in order of their plausibility to you?

2. Before Apollo 11, what did you regard as the major strengths and weaknesses (defects) of the various theories of lunar origin? Why?

3. After Apollo 11, what do you now regard as the most plausible theory? Least? Why? Rank the theories.

4. After Apollo 11, what do you now regard as the major strengths and weaknesses of the various theories?

5. In your opinion, are any of these theories now definitely ruled out? Why? By what specifically?

6. In your opinion, what would constitute final confirmatory evidence for a particular theory?

7. Can one begin to say what would constitute a crucial experiment that would decide between theories? Has such an experiment already been performed? Is one now in the making?

8. What in your opinion were the most significant and/or surprising results to emerge from Apollo 11?

9. Whom do you consider to be the outstanding theorists regarding the origin of the moon? What in your opinion distinguishes those scientists who formulate theories from those who do not?

10. Are there any scientists whom you regard as highly committed to, or even biased in favor of, their pet theories and hypotheses? Why? For which reasons?

11. Thus far, we've been talking about the negative aspects of bias. Are there any positive aspects to bias?

12. If commitment and even bias are so common in science, or at least much more common than the lay public has generally recognized, what happens to the notion of the objective, disinterested scientist in pursuit of truth? Is there such a thing as disinterested truth? How does scientific truth emerge?

13. What do you think of the following scientists as scientists, as people? Whom do you regard as the most similar to you, first in regard to holding the same general position on scientific issues and second in regard to the same general style of doing research? Whom would you regard as the most dissimilar to you in regard to these matters?

14. In your opinion, what is the significance of this study, that is, the study that I'm doing?

15. Which scientists would you recommend that I ask these same questions of next? For which reasons do you recommend that I see these persons?

Aside from their purely informative content, the first eight questions on the nature of the moon and the Apollo 11 mission were used as openers. They seemed to be sensible questions with which to begin the interviews. No less important, they were asked first because they are relatively neutral; they ask the individual scientists to start out by rating and evaluating physical theories, not their fellow scientists. This does not mean that these questions are not important in their own right, for they are. But just as important is that they served to get the interviews going by having the respondents talk first about an area in which supposedly they are the experts.

Although these first eight questions are important for background information, I do not give them here the primary attention that they could rightly command because, although they constitute a very high proportion of the initial set of interview questions, they do not constitute the major emphasis of the study. One of the major functions of the first eight questions was to act as relatively neutral lead-ins to the more sensitive and volatile questions about how the scientists perceived science and particularly their fellow scientists.

Like most of the questions in the section pertaining to science and scientists, question 9 served a variety of purposes. For one, it served as a lead-in to question 10. I felt it was desirable to turn the conversation gradually from the moon to science and to scientists. Question 10 was too volatile to approach directly. Besides, question 9 is important for background information. Like so many of the questions, 9 is so general that it allows considerable leeway in pursuing a variety of hidden and side topics. Depending on the respondent, one can follow an endless course of exploration on almost any one of these questions. Question 9 allows for the possibility of bringing out how the scientists would define the differences between, say, experimentalists and theoreticians and the special form those differences take in geology. The terms *experimentalist* and *theoretician* are so broad that it is unreasonable to expect that they would have the same meaning across all scientific fields.

To the best of my knowledge, the question (10) of commitment and bias in science has never been met head-on; it has never been, again to the best of my knowledge, asked systematically of a group of scientists. Prior to actually getting out and talking with

44

the scientists, I was somewhat apprehensive about question 10. It represented a gamble. If the scientists responded to it, they probably would respond to almost anything else I would like to ask. If not, the study might have been over before it really had had a chance to begin.

Questions 11 and 12 as they appear in the text, are different from the way they appeared in the initial set of questions. As they are worded, they represent a bit of looking ahead. As they appear in the text, they presuppose a positive response to question 10, which in fact was the case. That is, the overwhelming body of the respondents indicated that there was considerable commitment and bias in science. Even more important is that, within the context of the interviews, questions 10 and 11 were not loaded, as they may appear to be standing by themselves.

Finally, it should be noted that the preceding questions and categories are, of course, not independent. There is more than just a minimal degree of overlap between them. Thus, it is not surprising that the responses to the categories tended to overlap. What was surprising was the considerable degree to which the responses overlapped. In fact, the considerable extent to which the scientists themselves were unable to keep the categories strictly apart was one of the most interesting and significant findings to emerge from the first round. The scientists continually shifted back and forth between the categories. Time and time again it seemed tremendously difficult, if not impossible, for the scientists to discuss an issue supposedly pertaining to the moon solely as a physical object without involving, in some very intense and highly personal way, some colleague who was related to that issue and with whom they were embroiled in some dispute or in a competitive race. For this and other reasons that will be made clear, my efforts in the following chapter are mainly devoted to questions and issues pertaining to the nature of science and to the nature of scientists. Because this is a book which is primarily concerned with the nature of science and of scientists, the treatment of the scientific issues involved will necessarily be brief. I am concerned with the scientific ideas only to the extent that they allow us to follow the issues that are involved. This is not a book about geology or the moon. It is a book about scientists who happen for the most part to be geologists or geoscientists and whose object of inquiry is the moon. I defer a more extensive discussion of the attitudes of the scientists toward some of the scientific issues to a later chapter, where they can be dealt with on a systematic basis.

The first round of interviews was designed primarily to get an overall feel for the individual issues, so that on later rounds more systematic and quantitative ways could be constructed for assessing their structure. Above all, the first round was designed to explore the emotions connected with some potentially volatile issues that relate primarily to the scientists' perceptions of their peers. Without an appreciation of the intensity surrounding many of these issues, much of the purely statistical and quantitative data presented in later chapters is meaningless. The following chapter especially is designed to give the reader a feeling for the intense human emotions that lie behind the statistical data. Without this feel for the qualitative responses, it is unlikely that one can come to appreciate just how much science is a human drama performed by human players. It is also highly unlikely that, without having heard the scientists speaking for themselves, one can come to appreciate the pathos that is so much a part of scientific life.

PART II

Study of Moon Scientists

Chapter Three

Pathos of Science: Apollo 11

"In the first place,... you must keep the whole affair a profound secret. There is no more envious race of men than scientific discoverers. Many would start on the same journey. At all events, we will be the first in the field."

Jules Verne [451, p. 21]

In the preceding chapter I described the methodology underlying the first round interviews. This chapter presents some of the contents of those interviews and discusses some of their many critical implications. One of the primary concerns of this chapter is to show how the results of the interviews bear on the construction of a set of counter-norms that are contrary to the norms of science as they are currently conceived (see Chapter One). It is the concern of Chapter Seven to show how these same responses bear on the construction of a combined social psychological and philosophical theory of scientific objectivity. The results of this chapter make it imperative that we face squarely the issue as to how scientific objectivity is still possible given the results presented. However, it is best to defer this concern until after the results of the entire study have been reviewed.

The presentation of the interview responses is organized along the three broad categories into which the questions generally fell. The detailed plan of the chapter is as follows: First, a rather brief and general discussion of the various theories that have been proposed for the origin of the moon is presented. This discussion is intended mainly to give the reader a feel for the scientific issues so that he can follow the subsequent discussion. It is not intended to serve as a substitute for a serious and detailed discussion of the difficult technical issues involved; such a discussion is unfortunately beyond the scope and purpose of this book. For the reader who is interested in pursuing the technical points further, a number of popular texts [261, 281, 404] already exist that are excellently suited to the level and interests of the scientifically-minded layman.[1] Second, given these background ideas and again in accordance with the rationale of the last chapter, some of the responses to the broad category of questions pertaining to the moon are discussed briefly. Third, the main body of questions dealing with science and with scientists are discussed. Fourth, some very general and rather strong conclusions are drawn from the responses.

Origin of the Moon

Ralph B. Baldwin [29, p. 42—43] wrote: "We are thus left on the multipointed horns of a dilemma. There is no existing theory of the origin of the moon which gives a satisfactory explanation of the earth-moon system as we now know it. The moon is not an optical illusion or mirage. It exists and is associated with the earth. Before four and one-half billion years ago, the earth did not exist. Somehow in this period of time, the two bodies were formed and became partners. But how?" These comments by Baldwin sum up pretty well the situation before Apollo 11, and to an extent the situation even after Apollo 16. Although the Apollo data set some strong limits on a number of the theories for the origin of the moon, the data do not definitely rule out any of those theories.

As Levinson and Taylor [261][2] note, although there are a number of theories and countless variations on each theme, the multitude of theories fall into three main classes: (1) *fission from the earth* — the idea that the moon was once originally a part of the earth and subsequently broke away from it; (2) *capture by the earth* — the idea that the moon was once an independent body formed elsewhere, say, in a remote part of the solar system, then subsequently captured by the earth; and (3) *simultaneous formation along with the earth* — the idea that the moon was formed along with the earth as a double planet.

The class of hypotheses that are generally associated with fission stem from the idea first proposed in 1878 by George Darwin, the son of Charles Darwin. This is the idea that the moon was formed by a separation from the earth as the result of the action of solar tidal forces. The pattern of reasoning is as follows: If the mass of the moon and its orbital angular momentum were combined or concentrated into the mass and angular momentum of a liquid primitive earth, then the combined primitive earth-moon planet would rotate with the considerably faster period of revolution of about 4 hours. That is, the length of the day would be shortened, from 24 hours down to 4 hours. Because there are 2 solar tides every day (now 4 hours long), Darwin reasoned that the period of rotation of 4 hours might be close enough to the earth's fundamental frequency — which, for a combined earth-moon planet he calculated at about 2 hours — to allow a resonance action to be set up. Once set up, resonance would act indefinitely. The solar tides would act on the primitive earth until they built up to such a height that one of them would break off, and in this way the moon would be formed.

This hypothesis has a lot going for it, because it permits a marvelously simple explanation for a number of important facts. For one, it explains why the density of the earth's outer mantle and the overall density of the moon are so strikingly similar. For another, it helps to explain why the moon is unique among the satellites of all the other planets in the sense that our moon is so much larger relative to the mass of its parent body than are the moons of the other planets.

Unfortunately, like all the theories for lunar origin, the fission hypothesis has as many — perhaps more — weak points as it has strong points. Modern dynamical calculations show that Darwin's numbers were far too optimistic. The earth's fundamental frequency is considerably less than 2 hours and thus resonance would have been considerably more difficult to originally obtain. Further, the dynamics of deformation for forming a moon from a rapidly rotating spheroid are much more complicated than Darwin ever envisaged. For example, even if a molten blob were to break away from a rapidly rotating spheroid, if it broke away at any point less than approximately three earth radii (9,000 miles, a distance which is known as the Roche limit), the blob would be broken apart by the earth's gravitational forces. The Roche limit is the closest distance that a smaller body can approach a larger body before the smaller body is torn apart by the gravitational forces of the larger. For the masses and dimensions of the earth and moon, the Roche limit is approximately 9,000 miles. Not only are there other outstanding difficulties as well [see 29, p. 38], but since Apollo 11 new difficulties have arisen. The gist of these difficulties is based on chemical evidence [see 261, pp. 202—203].

Without getting embroiled in the fray between the objections and the counterrebuttals to the objections, at this point in the discussion one need merely note that the proponents of fission have not been silenced once and for all. There have been serious, well thought out, and ingenious mechanisms proposed to meet both the dynamical objections that were raised prior to Apollo 11 and the chemical objections that have been raised as a result of Apollo 11. As will be discussed shortly, although the respondents as a group did not rate fission very high, they did not rule it out completely either.

The capture hypothesis was suggested as a possibility primarily to account for two major facts. The first is that, relative to the plane of the earth's orbit around the sun, the moon's orbit is noticeably tilted. If the moon formed from the earth or in its immediate vicinity, it is not immediately apparent or easily explained why the moon's orbit should be inclined with respect to the plane of the earth's rotation about the sun. The second fact that the capture theory was suggested to account for is that the moon's bulk density is more representative of an object that was formed in the outer reaches of the solar system, specifically, the asteroid belt, than of a body that was formed in the immediate neighborhood of the earth. In short, according to the capture hypothesis, the moon was formed elsewhere and then subsequently captured from a highly eccentric orbit. As more than one commentator has remarked, this sounds a great deal like question-begging on the scale of the cosmos. It removes the origin of the moon to the outer reaches of the solar system without really accounting for it. No matter where the moon formed, we would like to know how and why.

Most theorists who have opposed or objected to capture have done so on the grounds that it is an extremely unlikely event; it is an event of extremely low intrinsic probability. Another way to put this is to observe that capture imposes some rather severe dynamical conditions and considerations. In order to account for capture, we are required to solve some rather formidable theoretical problems in celestial mechanics, and we are also obliged to indicate how these rather special and fortuitous circumstances came about physically. These essentially are some of the major criticisms that were lodged against capture prior to Apollo 11. After Apollo 11, the major evidence against capture, like that against fission, is chemical as well [see 261, p. 204].

The double planet hypothesis probably has as many or more variants and associated names with it as the two previous hypotheses combined. It has been variously referred to as the sediment-ring accretion hypothesis, the simultaneous formation hypothesis, the accretion hypothesis, or the condensation hypothesis. Although there are essential differences between the variants, the main idea is that of the simultaneous formation of the moon and the earth in the same neighborhood of one another. As first proposed by Opik [404], the theory suggests that the moon supposedly accreted from planetesimals orbiting the earth in rings much like the rings surrounding Saturn. Supporting evidence for this hypothesis is given by the rough outlines of elliptical craters on the very ancient lunar highlands. Supposedly planetesimals moving in the same orbit as an accreting moon would make impacts of this type in the lunar surface. Since the original proposal of Opik, this hypothesis as well has had to be revised to account for some of its more obvious discrepancies [see 261, p. 204].

These by no means exhaust the range of possibilities. Once the main forms have been enunciated, there seem to be as many additional possibilities as there are ways of combining them. One variant of capture is the notion that the moon as we know it accreted from many smaller moons, each of which was the result of a previous capture. This is known as the many-moon hypothesis. Certainly if there were many more smaller moons circling the primitive solar system, capture would become a much more likely process. Similarly, one can literally combine each of the other processes in an almost endless number of ways. However, no matter which process or combination we settle on, the situation still seems to call for the comment that was made a few years ago by two of the most distinguished investigators of the problem; namely, *it would seem that no method for the origin of the moon is possible; therefore, the moon cannot exist.* But as Baldwin and our senses remind us, the moon *is* there. Therefore, we are compelled, if not driven by our basic human nature, to give an explanation of why it is there.

The problem of the moon's origin is not typical of the vast majority of scientific problems. The problems that most scientists work on are probably nowhere near as ill-structured or as ill-defined; they do not permit as many irreconcilable and undecidable alternatives. In this sense, the problem of the moon's origin is not representative of scientific inquiry. However, in another sense, it is representative. Most of the great and interesting problems of science do not start out being well-defined but attain that hard-won status after a long history of repeated attack, if they ever do at all. Problems like the origin of the moon offer us the rare opportunity to study how problems that are essentially ill-structured become successively defined and converted into well-structured problems.[3] In this special sense, the moon's origin is representative of a special class and phase of all scientific problems, a phase that all problems must pass through if they are to become structured.

To some extent, problem types or states are fundamentally a reflection of the states of mind of the kinds of problem-solvers who work on them, rather than inherent characteristics of the problems themselves. For any problem there will always be those who strive to reduce it down to its "manageable and isolatable parts" and those who strive to highlight the essential "wholeness" of the problem [203–205, 310]. To a considerable extent, the choice of problem one works on is a reflection of one's basic style and one's nature rather than a reflection of the inherent nature of the problem. In this sense, the moon's origin is an excellent problem for study. Unlike the problems of fields that are already well structured and thus have already sorted out those who prefer to work on well-structured problems, problems that are as global and ill-defined as the moon's origin allow us to observe the sorting-out process. They allow us to observe how scientists characteristically sort themselves out with respect to the size of problem they prefer to work on. This is true even for a program like Apollo, where one of the main criteria for selection as a PI was the ability to formulate a well-defined experiment. Even in this case, we can still look for differences in the breadth of experiments proposed, or, because this is hard to do after the fact of selection, we can look for fundamental differences in attitudes with respect to the importance given to the breadth of interpretation of experimental data.

The question of the moon's origin hardly begins to exhaust the wealth of scientific issues, particularly those that emerged during the discussions. A number of additional issues will be taken up in later chapters.

Responses to Origin of the Moon

It is difficult to say precisely how the sample as a whole felt toward each of the various theories for the origin of the moon, either prior to Apollo 11 or just immediately after it. Although the interviews were a contributing factor to the lack of precision, the interviews cannot be held responsible as the main contributing factor. The interviews contributed by allowing a large number of the scientists to use the opening questions about the moon's origin as an opportunity or excuse to wander off into a discussion of numerous issues of particular interest to them. A number of the

54

scientists avoided the question of the moon's origin entirely. As a result, many opinions, sometimes contradictory, on many side issues, were expressed, and so a precise determination of who believed what is difficult although not impossible.

The main reasons for the lack of complete specificity of opinions were: (1) the fact that there was a significant number of scientists — *16 or 38 percent of the sample* — who either expressed no opinion at all with respect to the theories, or even stronger, in their own words were *"not interested enough in the theories"* to rank them; and (2) the fact that, of the remaining 26 or 62 percent of the sample who did express an opinion with respect to the theories, only 10 or 24 percent of the total sample ranked *all* of them explicitly. This latter fact is significant. The opening questions purposefully avoided mentioning any of the theories by name. One of the things I was most interested in observing was precisely how familiar each of the scientists was with the theories and whether they would explicitly consider or mention each of the theories by name in their rankings. Although the fact that only 24 percent of the sample explicitly ranked all of a limited number of theories is no proof by itself that the remaining scientists did not consider all of the theories, from the comments that were offered freely during the ranking procedure one can contend that in general there was a very selective consideration of theories.

Table 4 gives the rankings of the various theories in terms of their relative plausibility. The numbers in the column under the heading "Most I" indicate the number of times each theory was ranked most plausible; "Next II" indicates the number of times each theory was ranked second most plausible; "Least III" indicates the number of times a theory was ranked least plausible.[4] In the course of ranking theories, 18 of the scientists evaluated the theories in their own terms — for example, how probable or possible the various theories were. 12 of the 20 scientists who had previously ranked the theories in terms of their plausibility also evaluated them in terms of their possibility and probability (see Table 5). In addition, 6 scientists who had been either unable or unwilling to rank the theories were willing, able, or preferred to evaluate them in terms of probability or possibility.

Tables 4 and 5 reveal that, as a general trend, fission was rated the least plausible or possible of any of the theories. However, the kind of statistical test (a chi-square test) that is most appropriate for analyzing this kind of tabular data [288, pp. 230—231] indicates that the differences between the rankings and the evaluations

Table 4.

Rankings of Some Theories of Lunar Origin

Theory of origin	Plausibility ranking			Total
	Most I	Next II	Least III	
Fission	3	3	7	13
Capture	5	5	5	15
Simultaneous formation	5	3	0	8
Accretion from many moons	6	0	0	6
Total	19	11	12	42

Note: The table contains the rankings of 20 distinct individuals: 10 individuals gave 3 rankings; 2 gave 2 rankings; and 8 gave 1 ranking. A chi-square calculation for a 4×3 table gives a value of chi-square ~ 0.11; for 6 DF, chi-square $(0.05) = 12.59$; hence, the hypothesis that there is a difference between the rankings of the various theories is rejected [288, pp. 230-231].

across the theories are not statistically significant. And in fact quite the opposite is the case; the null hypothesis is significantly accepted in both situations; that is, the rankings are significantly the same for the various theories. Apollo 11 had not effected a significant sorting out of the theories. Later interview rounds will show a significant sorting (see Chapter Five).

The detailed comments and criticisms of the theories raised few points or objections that had not appeared in the literature prior to Apollo 11. This is directly attributable to the fact that, during the time period in which the interviews were conducted, it would have been almost impossible for any scientist to have assimilated the overwhelming data from Apollo 11. Also, prior to the first Apollo Lunar Science Conference held at Houston, Texas on January 5 through 8, 1970, there was a general moratorium and prohibition on the private discussion or public presentation of scientific results. Thus, the time for interacting with the data and with one's colleagues that is so necessary for changing attitudes had not been available. As a result, *few changes in attitude regarding the very special question of the moon's origin had been effected.* With attitudes concerning a host of more factual matters, the situation was vastly different. As a consequence, almost none of the scientists indicated that they had changed their opinions regarding the

Table 5.

Evaluations of Some Theories of Lunar Origin

Theories of origin	Possible	Impossible	Probable	Improbable	Combined[a] Possible probable	Impossible improbable
Fission	4	3	0	6	4	9
Capture	5	3	0	6	5	8[b]
Simultaneous formation	3	0	0	0	3	0
Accretion from many moons	1	0	0	0	1	0
Total	13	6	0	12	13	17

[a] The "Combined" table contains the evaluations of 18 distinct individuals; the opinions of 12 of these individuals have already been previously expressed in Table 1; thus, Table 2 adds the opinions of 6 new individuals. From Tables 1 and 2, the total number of scientists expressing themselves on the issues is 20 + 6, and therefore, there are 42 − 26 or 16 scientists not expressing themselves on the issues. X^2 for the "Combined" 2 X 4 table yields a value of $X^2 = 2.05$; for 3 DF, $X^2_{0.05} = 7.82$; hence, the hypothesis that there is a difference between the evaluations of the various theories is rejected [288, pp. 230-231].

[b] One scientist classified capture as both impossible and improbable; the "Combined" category has thus been reduced by 1 in order to reflect accurately the total numbers of scientists who felt capture was either impossible or improbable, but not both. In addition, one scientist characterized fission as both possible and improbable; another characterized capture in the same way; these evaluations have not been reduced because they are not necessarily contradictory opinions.

origin of the moon as a result of the Apollo 11 data, although roughly a quarter of the sample did indicate that factual constraints on the theories were beginning to emerge and form in their minds. Given that scientific ideas require time to change, I was in this sense quite fortunate. One of my greatest fears prior to the start of the interviews was that the scientific ideas would be changing so fast that initial attitudes would be lost. This fear turned out to be unfounded.

One of the most significant findings to emerge in response to the opening questions was not the number of scientists who held which position, although this information is important; nor was it even the fact that staunch proponents for each major theory for the origin of the moon were identified, although this, too, is important. The most significant finding was that there were, relatively, so many scientists who seemed so unconcerned with the question of the moon's origin. If, in addition to the 16 scientists who had no initial opinion or expressed outright little or no interest regarding the question of the moon's origin, we add the 5 scientists who — although they ranked the theories — indicated that they had no intense interest in them, then a full *21 scientists out of a total of 42, or 50 percent of the sample, expressed little or only moderate interest regarding the origin of the moon.* This is all the more striking, and calls for an explanation, because, of all the questions pertaining to the moon, perhaps nothing captures the interest and imagination of the layman more than the question of its origin. Further, given the fact that in virtually every culture known to man the moon has assumed a significance that is second almost to no other object [172], few things could appear to be more fundamental than the question of the origin of man's closest heavenly neighbor. More than one observer of the human psyche has noted that in a fundamental sense the quest for the nature of our own origin and our own identity is inextricably bound up with the quest for the origin of our lunar body.

But then this seems to assume that the interests of the scientist ought to match those of the layman. Perhaps the very thing that serves to differentiate the scientist from the layman is that the former has learned how to refine his interests. Thus, what appears as a lack of interest on the surface is really only a reflection of more detailed and deeply developed interests that extend below the surface of the subject. The scientist, it can be legitimately argued, has learned to forego his immediate interest in the ultimate questions for the less glamourous but nonetheless necessary process of compiling a long series of patiently accumulated facts in response to a series of small answerable questions. Unlike the layman, the scientist has learned that the ultimate questions are not answered all at once but only over the long haul through the pursuit of a large number of questions, each of which is individually answerable. Whatever the merits of this general line of reasoning, it is not the explanation for this particular situation.

Although there were very few scientists in the sample who had not had a history of some involvement with the moon or with lunar studies, it was generally felt by the respondents that there was a sizable number of scientists in the Apollo program whose interests in the moon were as permanent as the budget of NASA. In the words of a number of the respondents, "A lot of these guys came on board when the going was good and as soon as the money begins to dry up you'll see these guys get out faster than lightning" (see footnote 10). This statement may apply to some of the scientists in the sample, but to only a very few. It was more widely felt that those scientists who jumped on the bandwagon of the moon program were, for the most part, the kinds of scientists who were extremely wedded to a single experimental technique that could just as easily be applied to the analysis of moon rocks as to the analysis of something else. These scientists were not in the program purely for the money but primarily because the money allowed them to utilize, to develop, and to refine their special techniques. The interests of these scientists were characterized as narrow; they were described as the kinds of scientists who were likely to be interested in purely experimental questions, certainly not those who would be interested in the big questions. This does not mean that these scientists did not make important contributions to our understanding of the moon, but rather that they were not primarily interested in the study of the moon. It also does not mean that some of these scientists did not come to be interested in the moon.

What was imputed to be true of the general population of scientists was, to some extent, true of the sample. In strong terms, many of the scientists characterized themselves, and more often than not their colleagues, as being the same narrow kind of scientist. A statement that typifies the comment heard most frequently in this regard is the following: "You've got to realize that the *average scientist* [emphasis added] is not interested in these matters; the average scientist is not equipped either by virtue of his training or his natural inclination to get involved with or tackle these kinds of questions."

As strong as these remarks are and as much as they help to give some insight into the apparent lack of interest in the question of the moon's origin, there are even stronger reasons for that lack of interest. The gist of these reasons is most easily and best conveyed through a capsule statement that summarizes the feelings of a great many of the scientists. Although the following is not in its entirety a verbatim quote from any single one of the scientists, it embodies the direct statements and sentiments of a large number of individual scientists in the sample:

You've got to realize that we've lived with some of these theories for so long that they don't mean much to us anymore. To a large extent these are matters of almost pure speculation. We've heard the same old people spin out the same old cobwebs and speculations for years without adding much to them. We've tired of hearing the same old crap. Some of the people in this business have talked these ideas to death. You don't make progress by staying perpetually within the realm of theory. For the first time in history we have the hard data, the evidence to test a lot of these things. It's the data that's important now. This is the age — the day — of the experimentalist! Why do you [referring to me, the interviewer] keep pressing me for opinions on matters I don't think much about? You seem to imply that the only thing of interest is the origin of the moon. There are a whole range of things that are of interest to us, and you've only hit on one.

This statement begins to get at the real heart of the matter. The theories have been around for a long time, and no clear indication of their resolution seems to be in the offing. One important indication of this is that there were extremely few responses to the question of what would constitute a crucial test of the theories (question 7 in the preceding chapter). As one of the respondents put it, "If we had any clear idea of what constituted a crucial test, we would have done it by now." This, plus the fact that scientists are naturally disdainful of what cannot generally be put to some conceivable test, did not serve to endear the theories to the scientists. Time and time again, during the course of the interviews and at the various press conferences in Houston, I heard the statement, "This is the day of the experimentalist." I heard this so often that it could have served as the slogan for the entire scientific program.

Without my knowing it, the opening issues were themselves volatile, not because the origin of the moon is inherently a volatile issue, but because the scientists wanted to get on with the hard data and because there were strong feelings directed toward some of the scientists who were associated with the theories. It is putting it mildly to say that the behavior and personal characteristics of some of the scientists literally infuriated the general body of scientists in the sample.

I do not wish to create the false impression that none of the scientists in the sample were interested in the question of the moon's origin. This was not true. However, depending on what one takes as evidence for a belief or attitude, one-quarter to one-half of the sample expressed a lack of initial opinion or interest with respect to the theories. The fact that there was such a relatively large number of scientists who felt this way was the first behavioral indication of the personality differences among scientists that were later to emerge in more explicit and systematic ways. As one of the respondents put it, "Those questions are the concerns of the speculator types, the 'finger-painters-in-the-sky' types." If so, we might ask, what do the "finger-painters-in-sky" types look like? How prevalent are they? What other types are there? How do the various types as groups differ between and within one another?

Perhaps none of this is really surprising when we realize that, at any one point in time, only a relatively few individuals create and defend broad theories. Most scientists, it would seem, are concerned not with proposing and testing large-scale theories but with experimentally discovering and verifying small-scale theories or facts. In any case, the initial questions had more than served their intended function. I probably could not have designed more natural lead-ins to the questions on science and scientists if I had tried. Without any intention or design on my part, the initial questions naturally served to get the scientists to talk about their perceptions and feelings about one another. Those who are used to thinking of science in stereotyped terms, in which scientists are supposed to keep their personal feelings of their fellow scientists clearly distinct from their feelings towards the scientific issues, would do well to reflect on the consequences of this fact. I comment further on the implications of this later in the chapter.

Commitment and Bias in Science

commitment, n., a strong belief in search of supporting evidence and arguments such that it is not disturbed by its failure to find them.

bias, n., a strong belief unencumbered by supporting evidence or arguments and sustained by its firm opposition to all opposing beliefs.

There is really no natural division between the general body of questions pertaining to the scientists' perceptions of science (questions 11 and 12) and their perceptions of their fellow scientists (questions 9, 10, 13, and 15). Strong feelings were evinced about the nature of science because strong feelings were associated with many of the scientists connected with the lunar program.

This section discusses the general perception of science, especially the general nature and function of commitment and bias in science. The question of which scientists were perceived as most committed to their hypotheses (question 9) is discussed in a later section that deals with the questions pertaining to the scientists in this study. Logically there is no reason for discussing one set of questions before the other. However, because of the nature of the material, it proves easier to discuss the abstract notions of commitment and bias before the concrete personalities and the perceptions of the scientists are discussed.

The comments of this section cover the responses to questions 11 and 12, although these questions are so basic that they served to raise a number of fundamentally related issues to which the scientists then proceeded to respond. That the responses would necessarily involve a number of side issues was not unexpected, because the notions of commitment and bias (particularly their elimination and control) are so fundamental to the notion of science that they inherently raise many related issues. Thus, for example, to raise the question of a scientist's commitment to a particular point of view is to raise or necessarily involve the question of scientific objectivity, because these two notions stand in a direct relation to one another.

The results of this section are so much at variance with the common, stereotypical notion of science discussed in Chapter One that I have chosen to present a number of small excerpts from the interviews, plus a few longer passages, in order to give the reader a direct feeling for the intensity of the scientists' responses and the fact that that intensity was widespread. All the excerpts are either direct quotes or shortened paraphrases that are faithful to the original material. Although the excerpts represent a very small proportion of the total interviews, none of the excerpts are isolated sentiments that are unrepresentative of the general mood and responses to the questions. Each excerpt represents accurately the general feeling and position of each respondent toward the issue. As evidence for this assertion, I would stress that the interviews were not only given freely but that the statements were generally uttered with firm conviction and repeatedly over the course of the study. As even stronger supporting evidence, it should be noted that often it was not necessary for me to raise these questions explicitly. The scientists themselves raised many of these issues in the course of their open-ended remarks.

The scientists are referred to by fictitious names — Scientist Adams, Scientist Baker and so on. *The names have been randomly assigned and bear no relation whatsoever to the actual names of the respondents.* In order to further insure the anonymity of the scientists, more than one fictitious name will be used for the same

62

individual; very rarely will the same individual be designated by the same name. I would also add, largely for the benefit of the scientists who participated in this study, that in *no* case should the statements be attributed to any particular individual. Although I cannot prevent people from playing the game of guessing who said what, I can caution the reader that *in very few cases was something said that was not in effect uttered by at least one other individual.* Thus, for example, the mere extremeness of a response will not be enough to identify any particular respondent because so many of the respondents expressed extreme comments throughout the study.

One final word of initial commentary before turning to the responses and the main discussion: The terms *objectivity*, *bias*, and *commitment* each have many definitions and a long history of scientific usage. It is not my intention prior to the presentation of the responses to define these terms or to review the various meanings that these terms have assumed in the literature of science and philosophy. At this point, I feel it is much more important for the reader to get a direct feel for the responses than to be preoccupied with the nuances in meaning among various definitions of the same term. For the time being, the term *scientific bias* used in the *personal* sense means the tendency of a scientist to possess a degree of overcommitment to a scientific position, theory, or point of view that makes it extremely difficult for that scientist, or for anyone else, to modify his position by rational scientific arguments and/or evidence. In addition, scientific bias will also be used to imply a strong tendency on the part of a scientist to interpret evidence so that it fits his position, and conversely, the strong tendency to reject (most often unconsciously) evidence that goes against his position. In short, bias implies or connotes a very strong commitment to a position. Of the two terms, bias seems to be the more pejorative. That is, although all bias implies a strong commitment (not necessarily conscious), not all commitment seems to imply that one is biased. Of the two, commitment seems the more rational, bias the more irrational.

Those who are interested in pursuing the meaning of commitment further are referred to Charles A. Kiesler's interesting monograph, *The Psychology of Commitment* [226], in which Kiesler reports in detail on a fascinating series of simple experiments that attempt to get at the meanings of commitment in behavioral terms. Kiesler offers the following working definition of commitment: "For purposes of discussion, commitment shall be taken to mean the *pledging or binding of the individual to behavioral acts"* [226, p. 30]. Kiesler argues that this definition implies that the greater the degree of commitment an individual has to a position, the harder it should be to shift him out of that position: "The effect of commitment is to make [a behavioral] act less changeable" [226, p. 31]. I will not pause to pursue here all the ramifications of this approach to commitment. It should be noted, however, that the general definitions of bias and commitment given above could be made more operational if one so desired, and that the definitions could be shown to square with the existing behavioral data.

Each round of interviews contained at least one question that served to provoke a common if not almost universal response from the sample. In the language of statistics, the variance was the tightest for this question. For round I, it was question 12 — the notion of the objective scientist engaged in the disinterested pursuit of truth — that provoked this kind of response. All of the 42 scientists interviewed indicated in one way or another that they thought the notion of the objective, disinterested scientist was naive. The vocal and facial expressions that accompanied the verbal responses were extremely revealing and important. They ranged all the way from mild humor and guffaws to extreme annoyance and clear expressions of anger. The respondents felt that the only people who took the idea of the objective, disinterested scientist literally and seriously were the general public or beginning science students. Certainly no working scientist, to quote one of the respondents, "believed in that simple-minded nonsense." Because they actually did science and because they had to live with the day-to-day behavior of some of their more extreme colleagues, they knew better.

Because of their direct contact with the actual workings of science, and because of their strong feelings toward some of the scientists described later in this chapter, it is not surprising that the scientists were disdainful of the Storybook account as an accurate descriptive account of science. What is so surprising is the extent to which many of the comments revealed that, even as a prescriptive account, the Storybook version of science was considered naive and found wanting. The scientists not only felt that the Storybook account failed to describe the actual workings of science but even rejected it as an ideal that one should aim toward. Below are quotes from some of the respondents:

64

Scientist Adams: Bias has a role to play in science and it serves it well. Part of the business [of science] is to sift the evidence and to come to the right conclusion, and to do this you must have people who argue for both sides of the evidence. This is the only way in which we can straighten the situation out. I wouldn't like scientists to be without bias since a lot of the sides of the argument would never be presented. We must be emotionally committed to the things we do energetically. No one is able to do anything with liberal energy if there is no emotion connected with it.

Scientist Baker: Every scientist feels committed to put his work forward in the best possible light. It's only natural. We'd be funny people if we didn't. The idea of the dispassionate, disinterested observer just isn't so. The more you work on a subject the deeper the commitment becomes. It is important for scientists to take strong stands. Even a wrong theory or hypothesis has tremendous value to science, especially if it is put forward provocatively. It stimulates people. It makes them so mad they get to work to disprove it. I once set out to disprove Jones and ended up validating him.

Scientist Case: You can't understand science in terms of the Boy Scout articles that appear in the journals. It's better to approach it as the chairman of a political party.

Scientist Davis: I can't find any fundamental difference between the scientific method and the procedures for making progress in business and the arts. Where the evidence is not clear-cut, people reach out and grab for straws. They are extremely subjective and intuitive. They invent elaborate analytic justifications to convince their colleagues. Most scientists would say science has firmer criteria for the judgment of quality than business, but I disagree with this. Science is more than a Baconian activity. Bacon did irreparable harm to science. It is the selection of the facts that matters. Science is not an endless series of facts. Only the most pedestrian idiots, even though they may have PhDs, go about acquiring more facts, doing more measurements. It's the element of choice. You choose only by drawing on the most intuitive and most deeply buried recesses of your mind. The false myth of objectivity actually covers up your tracks of discovery. As a result one never writes it up the way one actually did it.

Scientist Edwards: The uninvolved, unemotional scientist is just as much a fiction as the mad scientist who will destroy the world for knowledge. Most of the scientists I know have theories and are looking for data to support them; they're not sorting impersonally through the data looking for a theory to fit the data. You've got to make a clear distinction between not being objective and cheating. A good scientist will not be above changing his theory if he gets a preponderance of evidence that doesn't support it, but basically he's looking to defend it. Without commitment one wouldn't have the energy, the drive to press forward, sometimes against extremely difficult odds. Trying to collect data on the moon is not the easiest thing in the world [this may be the biggest understatement of the entire study]. There are not only physical problems but there are bureaucratic problems as well to fight. You don't consciously falsify evidence in science, but you put less priority on a piece of data that goes against you. No reputable scientist does this consciously, but you do it subconsciously.

Scientist Ford: In order to be heard you have to overcommit yourself. There's so much stuff if you don't speak out you won't get heard, but you can't be too outrageous or you'll get labeled as a crackpot; you have to be just outrageous enough. If you have an idea, you have to pursue it as hard as you can. You have to ride a horse to the end of the road.[5]

Scientist Gage: The notion of the disinterested scientist is really a myth that deserves to be put to rest. Those scientists who are committed to the myth have an intensity of commitment which belies the myth. Those scientists who are the movers are not indifferent. One has to be deeply involved in order to do good work. There is the danger that the bolder the scientist is with regard to the nature of his ideas, the more likely he is to become strongly committed to his ideas. I don't think we have good science because we have adversaries but that it is in the attempt to follow the creed and the ritual of scientific method that the scientist finds himself unconsciously thrust in the role of an adversary.

Before commenting on the implications of these excerpts I would like to present two considerably longer passages that give much more of the total flavor of the interviews. These passages are not only interesting and significant for what they say about the nature of scientific inquiry but also for what they reveal about what science as an institution and as a way of life does to and demands of some of its practitioners. Because of their pointedness, I caution the reader that these are not isolated examples:

Scientist Logan: When I first came into this field [lunar science] I was dismayed at the complete lack of objectivity. I felt this was more characteristic of this field of inquiry than other fields, but when I talked to people in other fields, I found it was substantial there too, although perhaps not as much. People become strong advocates, and more than that, biased by their position. People have a selective memory and tend to remember their old papers better and longer than other people's papers. Thus one can read about an idea, forget it, and rediscover it, believing he is the first to have the idea. Also, sometimes people are reluctant to accept evidence as final and definitive. They have reservations, sometimes justified, about the quality of the data or the interpretation. In the controversy, what is unshakable evidence to one is just a flimsy argument to another. Sometimes one despairs of ever reaching any kind of universal agreement on a subject, on what's admissible evidence, and definitive evidence, and on what can safely be disregarded. A person should stick to his guns for a while, but there comes a point when the evidence is overwhelming against him, when it would be better for everybody if he admitted he was wrong. The crucial test is whether people quote your work and accept it. I have no real complaints. There is a natural selection process. There are times when I think I have received a little less credit than I've deserved. But I've established a reasonably good reputation. When I publish a paper, people read it and take it seriously. They may not believe every word of it and they may design experiments to prove me wrong, but at least they seem to pay attention to it, and that's all you can expect. Each paper is a stepping stone. Other people can improve on it. I am quite satisfied.

When asked how committed he is to his own ideas, Logan said:

There are quite a few documented cases where I have criticized my own work in print and have been fairly severe with myself. When there were weaknesses in my earlier theories, I would point this out. I think I have two kinds of reputations. Some people say I'm one of the most argumentative and prejudiced people in the business. I've disagreed with Smith on the question of phenomenon P. Smith would thus say that I was prejudiced. However, Smith hasn't read my papers in recent years; he hasn't read them carefully. He misquotes me. It's quite obvious that he hasn't even attempted to grasp what I am saying, yet he quotes me. Perhaps I have been unduly fierce and emotional in controversies, especially when I think I've been falsely accused of something. The question of phenomenon P was very bitter. It settled in my favor but left some scars. Perhaps I was a little too ferocious in some verbal confrontations. If I were to do it again I would be more mellow. Wilson brought out the worst in me. Wilson has a very limited theoretical insight. He is so limited that there are real communication problems at times. Wilson was so very sure he was right. Wilson and I were literally exasperated with each other. A number of people sat on the sidelines and enjoyed watching the fur fly. Some people will thus remember me as a person that gets into arguments, who proves that other people are wrong. On the other hand, some people will point out some ideas that I've contributed to the field, theories which have stood up well. Some of these people are PIs.

In response to the question regarding who would vouch for his work, Logan went on to list the names of about 15 PIs.

Scientist Meade: It was kind of fun at Houston [at the first Lunar Science Conference held on January 5—8, 1970, to discuss the results of Apollo 11] to hear one or two talks where a guy went off on big theories; your adrenalin really gets going; you ask how a guy can believe that crap. It gets dangerous to get too much of this as this is what's automatically picked up by the press. It's what the average guy can understand. This makes the nontheoretician rather paranoid about the whole business — the publicity, the fanfare, the grandiose theories. The guys who try to interpret with a little bit of balance are not understood. I can write a paper and send it to Jones and to Smith, the director [of Meade's organization] around here and they won't understand much of it; they won't approve it, but will pass it on. On the other hand, they will understand Hall's paper on a theory because he will leave out most of the data and paint a big picture without the details and is therefore very proper. People know his theories. [When I asked if there was a good theorist who doesn't leave out data, Meade replied that he didn't know one.] We all get rather paranoid [this is repeated a number of times]. Scientists are probably as disturbed a people as any. They get paranoid about their role, their work; they feel no one understands them, that others screw them. This is terrible [this too is repeated a number of times]. Scientists are terribly competitive due to this [repeated a number of times]. [The kinds of scientists who work on phenomenon Q] are probably the worst of all, a really competitive backbiting group.

Meade has become rather bitter and cynical about the theoreticians.

A theoretician is someone who doesn't understand the data, but will understand the limitations and assumptions in the data. Smith was going pretty far with the data in Houston; he made statements that had more than one interpretation. There were assumptions involved that Smith hadn't expressed, and he probably doesn't understand it. I didn't say this to him; I chickened out. There were all theoreticians up there on the panel [of discussants] and I didn't want to get stoned by them. I find it very hard to keep up in a couple of very small fields; it's next to impossible. I wonder how people get their data, how good it is, what their assumptions are. Then I run across the theoreticians who try to portray all of geology, celestial mechanics, all these fields, and throw them into large theories. I find it difficult to understand what they are talking about. When I watch them operate in my small fields, I see that they don't understand them. It seems to me that those who keep hammering on something, who take very strong positions, are those who get something across. If you give a fifteen minute paper at AGU [American Geophysical Union] and then go on to something else, you aren't going to make much of a splash. The big people give the same talk at 6 or 8 different places, adding only enough data so it's publishable. They push very hard for an idea. This is more than just the theoreticians versus the experimentalists. Both have to sell a product. It's like a life style. You develop certain beliefs in life, and believe yours are better than the other person's. If you do a piece of work you become convinced that it says something, and believe in what you can interpret. There is nothing wrong with being committed this way. You try to sell the possibility you think most likely. The dishonesty comes in when you completely abuse your position and use it to destroy the opposition. This happens quite often.

Meade has seen Park [a very prominent scientist] do this.

Human nature is human nature. You're going to pull in front of a guy, if it gets you something. Most people don't think about it; they are often quite cruel in an offhand way. Once at a conference someone asked Park a question at odds with Park's beliefs, and Park cut him to pieces with a joke. The audience laughed and the guy was dead, and Park carried on. You can see this happen all the time. Everybody loads arguments on his side. People aren't really trying to reach an agreement as much as convince the other guy of what they have to say [this is repeated]. They don't want to discuss, teach, or learn. In most arguments both sides push their point, and say the other son-of-a-bitch doesn't know what he's talking about. The same is true in science.

Meade likes to think that "the training involved in becoming a scientist controls the conscious manipulation [that is, cheating] of data." I then said I wasn't talking so much about the conscious manipulation of data as much as the unconscious selection or manipulation. Meade then said,

You can't influence the collection of data. You can influence the experiments done. I can prove that phenomenon P is the same as phenomenon Q if I do such and such experiments. Your bias determines what experiments you do. You have to do this. There's too much to do. You can't work in a field without reaching conclusions about it. Your data tells you something, so you are going to be biased [sic]. *I don't think this is necessarily bad. I only object to it when it becomes an obsession* [emphasis added]. I'd be embarrassed if my numbers were wrong, if I made a bad error in my measurements [this is repeated]. I've been trained more along the lines of collecting data. The original sin in my field is: You Will Not Publish Bad Data! [sic] Theories come and go, but the data will stay in the literature forever, and people will be hanging theories on that stuff for 100 years or more [this is repeated]. When someone publishes in my field, the first thing I look at is the data; I feel good if it agrees with mine. If his theory and mine don't agree, I might get a little upset but not as much as I would if the numbers didn't match. Luckily this has never happened.

When I asked Meade how he happened to become an experimentalist as opposed to a theoretician, Meade said,

Jesus, I don't know; I don't know. I guess I always had a preference for experimental work. I've never thought about it. Maybe it's due to where I went to school, who I went to school with. I did my grad work at [one of the major universities]. What you end up doing is partly a matter of very small circumstances that you really don't recognize at the time, which influence your whole life [this is repeated]. At one time I thought I would become a general descriptive geologist. I was advised I lacked the math and the physics to be a geophysicist. I wanted to be a general geologist anyway, to look at things. But, geology isn't run that way anymore. It's getting more and more quantitative. So I went to [the university], one of the strong schools in quantitative geology. Then I couldn't get a job. The X business in geology was in a depression so I stayed in school, was offered an assistantship — where I ended up depended on which professor approached me (another small circumstance). I was approached by a guy who had a project on phenomenon P and so I did my thesis on it. And here I am today — a failure [laugh]. I don't think I would be any happier doing field geology. I think I'm satisfied with the progress of my career. I don't know. I probably sound like I'm paranoid. Hell, maybe I am. I don't know. I take things easy. I don't really worry about this. Everybody does, some more than others.

In response to question 14, the significance of this study, Meade replied,

There is a value in convincing the public that scientists are like everyone else [this was repeated] — same greed, same drive for success, power. When a scientist says it, boy that's supposed to be it. You always want to be aware that especially when a scientist talks about chlorides in water, pollution, cigarettes, people will use these things for their own end. The days of the scientist being god are coming to an end. I suppose this study would be good for young science students to read. I've often thought about developing a talk on the politics of geology for students. Sometimes I get cynical enough to want to tell grad students what they're really getting into — a competitive world where success is not necessarily dependent on how good a scientist you are. Some are very successful in pure science but it really isn't pure; nobody is pure. I don't think your study will be useful in changing all this. It would be interesting for students to hear more about the politics of science. It would help them to be better scientists from the start. People want to sell their point of view, beat down the other guy because it means more glory, more ego satisfaction, more money. A scientist has the same wolves biting at his rear end as everybody else, and the same internal wolves biting at him, his ego, his own worth. If you had scientists who were completely psychologically normal, balanced human beings, they wouldn't be scientists. They would be objective then, but would there be real progress? There is an ego drive in Watson [the Watson of *The Double Helix*, 458] which takes away from his being normal. Watson worked to beat somebody in the worst sort of way; there was probably a real psychological need in him to prove he was worthy. The worst thing you can do for your kid is to make him a middle-class, well-balanced kid with no drive. Maybe the best thing you can do is raise a kid to be very deprived, psychologically and materially, so that he has tremendous drives for success, will trample over people and work like hell.

Given the extreme intensity and obvious poignancy of the immediately preceding remarks, I should explain why I have chosen to cite them. Certainly it is not out of any perverse desire to single out or to play with the feelings and emotions of a particular individual (whether he is nameless or not). Rather, it is for the prime purpose of illustrating vividly the inner and often extreme emotions that are connected with the doing of science. It is precisely these kinds of strong feelings that the interviews uncovered and which the Storybook account sloughs over or sees only in a negative light. Further, it is precisely these kinds of feelings which underlie the sociologists' conception of the social norms of science and which I believe they have not taken adequately into account. And so I have purposefully chosen to cite a real interview rather than present a hypothetical composite portrait of the interviews as a whole.

Finally, I want to emphasize again that, as extreme as the immediately preceding passages appear, they are not atypical of the interviews as a whole. In nearly every interview there are passages which match the preceding in intensity. I am not saying that all the interviews contain the same ideas or the same sustained degree of emotionalism but that there is a strong degree of emotionalism and poignancy in every one of them.[6]

The picture of science that emerges from the interviews is greatly at odds with the Storybook image. It differs with respect to the gross features and with respect to the substantive details as well. If we merely concentrate on the particular notion of commitment, then the excerpts reveal that, for the sample, the system functions nowhere near what the Storybook would prescribe. Hence, the Storybook is not an accurate descriptive account of science, or at least not for this sample. However, once this is acknowledged, there is much less agreement with respect to the implications of this fact. For example, there were scientists such as the one quoted below who, although they freely acknowledged that there was considerable commitment and bias in science, were less than enthusiastic about it and saw little positive value in it:

Scientist Nolan: It's the scientist's duty to convince himself, not others. Being strongly committed to the outcome of an experiment while conducting it is a very dangerous thing, and many scientists consciously or unconsciously do this; this is disastrous in science. It's the lesser scientist who gets committed." [Contrast this with the statements of Scientist Gage, who believes that it is the greater or bolder scientists who become committed.] The system rewards the spectacular. A whole lot of scientists are on the negative side of the bull shit balance.

Scientist Oaks: I don't believe it is the job of the scientist to argue most persuasively for his own hypotheses. The man who doesn't admit to negative evidence probably doesn't possess a scientific mind.

Nevertheless, the number of scientists who saw a positive side to commitment and bias was far greater. Instead of perceiving commitment and bias as inherently harmful to scientific inquiry and to scientific objectivity, a surprisingly strong majority of the scientists were able to spell out the positive benefits. If science was not necessarily better off for commitment and bias, it would not necessarily be the better off without them. This is not to say that the responses constituted an unqualified endorsement for commitment or bias. Nothing could be further from the comments or from the truth. Indeed, it is in the qualifications and restraints on commitment and bias that their positive function in science can be seen.

To pursue these points further we must consider some of the detailed points in the excerpts. At a first glance, it would appear that the excerpts contain some aspect of nearly everyone's conception of science. And in a sense this is true. There is no logical reason to adopt one view of science to the exclusion of others. For example, a falsificationist can find in the comments of Scientist Baker support for his conception of science — the idea that science advances through the process of proposing bold hypotheses and then by doing everything in one's power to discredit those hypotheses. However much I agree that the evidence supports a falsificationist interpretation, I don't think it supports the kind of falsification which makes a firm distinction between the context of discovery and the context of justification [347; see also Chapter One]. The contexts of discovery and of justification are at least bound together through the intense emotional commitment that requires a scientist first to discover his ideas and then to test them. The comments lend support to the notion that the process of testing or justification is intimately bound to the process of discovery through the intense psychological emotions needed to keep the whole process of scientific inquiry alive. The distinction between the contexts of discovery and justification promotes a separable account of science that the interview materials do not support.

As I read the excerpts and the broader materials uncovered by the study, I am impressed by the fact that science does not advance through the single efforts of individuals each dispassionately and logically testing their own ideas. Rather, it advances through a heated adversary process, which is fundamentally social, wherein one man tests his discoveries against the discoveries of another. Psychological energy and commitment[7] infuses the whole process to such a degree that it is foolish to say that scientific inquiry naturally exhibits a clear-cut dividing line, between individual scientists or between the contexts of discovery and of justification. This is the basis for the strong contention, made in Chapter One, that the Reichenbach distinction between discovery and justification rests on a naive psychology of science. To remove commitment and even bias from scientific inquiry may be to remove one of the strongest sustaining forces both for the discovery of scientific ideas and for their subsequent testing. If we have a right to demand of any account of science that it bear at least some relation to actual practice [207, 284], then I believe that we have a right to question the fruitfulness, if not the validity, of any account that is founded on a rigid distinction between the contexts of discovery and testing.

Alternate Norms in Science

Previous studies in the sociology of science have argued for emotional disinterestedness as a necessary condition for the development of science. The comments uncovered by this study indicate that there is a sense in which emotional commitment can also be viewed as a necessary condition for the development of science (see footnote 7). In this regard the comments of Scientist Gage are particularly important. They indicate that if disinterestedness deserves to be considered as a norm of science, then commitment and bias play a positive role in science and also deserve to be elevated to the regal status of norms of science. Further, the comments of Scientist Gage indicate that the new norm of commitment arises in part as a legitimate response to the old or conventional norm of disinterestedness. In this regard, the interview materials strongly support Merton's ideas on sociological ambivalence [298]. It would appear, in other words, that the case can be generalized beyond the particular norms of disinterestedness and commitment. For every one of the original norms of science, a case can be made for the existence of a corresponding counternorm. To see this, it is necessary to consider some of the additional interview materials which bear on some of the other conventional norms of science. Consider the conventional norms of universalism and community:

Universalism: in science all men have morally equal claims to the discovery and possession of rational knowledge.
Community: private property rights are reduced to credit for priority of discovery; secrecy, thus becomes an immoral act [468, p. 54].

On the face of it, it would seem absurd to contend that there could be some norms on the opposing side which could serve some positive function in science. The idea that there might be such norms came out during one of the interviews. Completely on his own and without any prompting from me, Scientist Park used the question of how he felt about a list of selected scientists in the field (question 14) to launch into an hour and a quarter discussion of who the biggest crooks in the field are. In response to my list, he produced his own list of scientists in the field and then proceeded to say how honest each is, defining along the way the particular kind of scientific thievery a particular individual engaged in. He also later supplied me with personal letters that were supposed to document many assertions. He also praised many of the scientists for their scrupulous honesty. Scientist Park almost exclusively, if not obsessively, uses the concept of honesty to evaluate his fellow scientists:

> Scientist Park: Jones is 95 percent politician; he would tend to be *honest* [emphasis added] if it didn't cost him anything, but if he can gain anything politically by being *dishonest* [emphasis added] he wouldn't hesitate. He wouldn't fudge data, but if he thinks he can gain something by giving Smith a lot of support he wouldn't hesitate. Zimmer is the most *dishonest* [emphasis added] in the whole field; Zimmer *steals* [emphasis added] people's work and publishes it if he can. He won't acknowledge the work of others. Neither will Wade. It's a political thing. They don't reference the work of anyone except those in their own group or people that get along with them. The well-known people who are *dishonest* [emphasis added] are very small in number. This includes Jones and Smith, but not the people who work for them. Two groups in the field told Jones about my discovery of Q. Jones tried to duplicate it, but couldn't. In Smith's case it's flagrant. He *stole* [emphasis added] and redid some of my work. *The majority of people are honest* [emphasis added]. Wilson published some of my work he redid in a foreign journal. He claimed he didn't know about it. I never confronted Wilson because it's so petty. He doesn't feel any embarrassment. *A lot of them don't even know they do it* [emphasis added]. I wrote Tate about the work he was doing with regard to phenomenon X; they filed the letter away and then acted like they had the original idea. I couldn't say I suggested the idea to them. Maybe the people working with Tate didn't know it. This is a borderline case. You have to be mature and go on in spite of it. There is *dishonesty* [emphasis added] everywhere.

If one considers only these comments by Scientist Park, he would seem to support the idea of universalism and community as ideal norms. Universalism and community may not be the actual norms-in-use, but the comments of Scientist Park seem to lend

74

support to the idea that they should be. However, Scientist Park made some additional comments that support quite an opposite interpretation. One of the most significant points he made was about timing (when it was that people began to steal from him) and hence when he first began to be aware of stealing in science:

> Scientist Park: It was only when I began to do something significant and important that people began to *steal* [emphasis added] from me. When I began to manage a big research program and all the big, important people began to visit me, they would rush home and try to outdo our results. *You know you're doing something significant when people want to steal it* [emphasis added]. It was only when I began to do something significant that I became aware of all this, what it's really like in science.

Given this real or imagined account of the behavior of scientists, it is obviously advantageous to scientists for them to operate with a degree of secrecy. If science were in actuality the free and open society that the norms of universalism and community make it out to be, then it would be in the interest of scientists to act in accordance with these norms. But science is not, or at least there is some evidence that it is not the totally free and open society it has been supposed to be. The error lies with confusing the norms of science as ideal prescriptive rules with the actual operating rules. This raises the possibility that secrecy may be the inevitable response to an unpleasant fact of scientific life (stealing). But acknowledgment of an unpleasant fact should not be sufficient cause for elevating that fact into an ideal norm. To be forced to accept an unpleasantry is one thing; turning it into an ideal is another. And I would agree with this line of reasoning if, apart from stealing, there were no positive aspects to the idea of secrecy.

As an opposing ideal norm to universalism and community, secrecy serves a number of positive functions in science. First, rather than being detrimental to the stability and progress of science, secrecy can further the progress of science. If there were no protective actions or countermeasures at its disposal, then the social system of science would be continually racked by the kinds of open internal disputes for priority that have been so aptly described by Merton [294—296]. Without some device such as secrecy, science would degenerate into a state of continual warfare. This contention becomes all the more worthy of serious consideration once we acknowledge, as so many of the respondents did, that the *out and out* stealing of ideas in science is extremely rare.[8] Thus it is not so much in response to the extreme practice of outright stealing that secrecy is postulated as a norm, but rather it is in response to the acknowledged and much more prevalent practice of *unconscious and unintended appropriation of ideas.* Recall the comment of Scientist Park: "A lot of them don't even know they do it." Secrecy is a natural protective response to the appropriation of ideas, whether that appropriation is real or only imagined.

A second — and perhaps the most interesting and important — function that secrecy serves is not as a reactive control after stealing has occurred but as a before-the-fact indicator (or acknowledgment) to oneself and to others that one has something in the works worth protecting. In this regard, some stealing or appropriation may not only be tolerable but even beneficial, as long as it does not reach epidemic or uncontrollable proportions. Although I would be quite reluctant to make stealing or appropriation into an explicit norm, as I am willing to do with secrecy, I would acknowledge that in a perverse sense even stealing and appropriation can serve some positive function. Perverse and dangerous as they are, stealing and appropriation may be among the most important ways of informing a scientist and his peers of the measure of significance and importance of his work.

In science, statistical significance is one measure of the importance of a work. Perhaps the sociological test of the real significance of a scientist's work is whether it is worth stealing. There has long been an unwritten (but not unspoken) rule of science: Do not divulge what you are up to until you are 99 percent sure that you have beaten the competition in the race to print. (See note 15 for one scientist's expression of this sentiment.)

I am not suggesting that secrecy be made an unrestricted ideal of scientific life. If science were exclusively to follow the norms of commitment and secrecy, science would run the danger of degenerating into a complete subjectivity in the case of commitment and into a complete solipsism in the case of secrecy. If science were exclusively founded on secrecy, I doubt that there could be science in the sense we know it. Public communication, sharing, and testing of ideas would all but vanish. But the key word is *exclusively.* For if science were also exclusively founded on the norms of disinterestedness, universalism, and community, then I doubt that the science we know could have arisen.

My point is precisely that each of the norms of science is restrained and that if any one of them were to operate in an unconstrained manner there would be chaos. But it is also my point that for every one of the conventional norms of science there are good reasons, consisting of arguments plus empirical evidence, for seeking to establish counter-norms. It is the norms on both sides which restrain one another and not just the norms on one side or the other. More will be said in Chapter Seven about how the norms might restrain one another. Specifically, I discuss how scientific objectivity might still be possible from a process of sharply opposing norms.

I am not contending that the counter-norms are any more valuable than the conventional norms. However they are formulated, any set of opposing norms will merely be one out of possibly infinite sets of contraries. The key word here is *polarity*. I have been too deeply impressed and influenced by the recent works on the polarities of the human psyche [53, 187, 214—216, 273] not to expect those polarities to manifest themselves in all of man's institutions and activities. I believe that any sensible account of science must not only account for the rational in man's nature but also for the irrational and nonrational [104]. This is not to condone stealing and secrecy in all their hideous shapes and forms, nor all the other aberrations of human behavior, but merely to acknowledge that for too long we have avoided the fundamental question. The fundamental question is not how science is possible because it is inherently pure or because it inherently operates by excluding all the aberrant features of human behavior, but how science is possible because it includes the aberrant as well as the rational. Has science learned to profit from the aberrant, not by excluding it all together but by learning how to control it?

I should like also to say a brief word about the data on which the strong proposals of this section are based. Again, I am not saying that the strong comments of Scientist Park are typical of the respondents. Although the topic of stealing was broached with many of the scientists, most of them indicated that it was not a serious problem for science. Thus, I am not basing my proposal for the serious consideration (postulation) of a norm like secrecy on the representativeness of Scientist Park's remarks but on the fact that he gives us a glimpse, as extreme as it may be, into the nature and possible function of secrecy in science. (See footnote 8 and the next chapter, which indicate that stealing may be more prevalent than we have thought.) In dealing with a little discussed and hidden aspect of science, we may have no recourse, at least not in the early recorded study of a subject, but to turn to those few who are more than willing to talk about a taboo topic precisely because they have had an extreme reaction to an extreme experience.

However, I would also add some additional information that helps to soften the apparent extremeness and isolation of Scientist Park's remarks. As delicately as it could be done and in the context of the inevitable natural opportunities that presented themselves during the course of the interviews, I very indirectly raised with some of Scientist Park's colleagues the notion that some of Park's ideas might have been stolen. Not all of them agreed with his strong reactions, but all seemed to agree that there was some basis for his allegations, and there were even a few of his colleagues who thought that his strong feelings were justified, that Scientist Park had been seriously wronged.[9] As much as possible I tried to institute cross-checks throughout the study. Without explicitly mentioning names or divulging the content of any previous interviews, I tried to get the perceptions of those on the opposite side of an issue or critical incident. This helps to provide a check on the perceptions, but I cannot contend that it provides a perfect check. Ultimately, I do not think there is any firm way of getting around the fact that fundamentally we are dealing with the beliefs of scientists. I went through a number of the personal letters and documents of the respondents, many of which tended to substantiate their perceptions of many of the incidents. But I do not see how one could prove many of the versions reported to me.

This book is about what scientists believe and are willing to believe about science and about one another. Substantiation of many of the points would be helpful and interesting, but lack of it need not detain us. The beliefs of scientists, whether accurate or not, have consequences for the conceptualization of science. As W.I. Thomas put it, "If men define situations as real, they are real in their consequences." Because they are scientists, what they think or believe about science is important.

Below is a proposed set of counter-norms. I regard this as a tentative effort in this area, and I hope it will encourage others to formulate additional sets. I realize only too well that the norms on the opposing side are no more free of internal contradiction, competition, or ambivalence than the norms on the conventional side. I also realize that many of the counter-norms are just as high sounding and pompous as the ones to which they are opposed. This was not done without conscious intent.

A Dialectic Between the Conventional Norms of Science and a Proposed Set of Counter-norms

Conventional Norms	Counter Norms
(1) *Faith in rationality.*	(1) *Faith in rationality and nonrationality* [104].
(2) *Emotional neutrality* as an instrumental condition for the achievement of rationality.	(2) *Emotional commitment* as an instrumental condition for the achievement of rationality.
(3) *Universalism*: In science all have morally equal claims to the discovery and possession of rational knowledge.	(3) *Particularism*: In science some men have special claims to the discovery and possession of rational knowledge.[10]
(4) *Individualism* (which expresses itself in science particularly as anti-authoritarianism).	(4) *Societalism* (which expresses itself in science in contrast to the lawlessness and chaos of anarchism).
(5) *Community*: Private property rights are reduced to credit for priority of discovery; secrecy thus becomes an immoral act.	(5) *Solitariness*: Private property rights are expanded to include control over the disposition of one's discoveries; secrecy thus becomes a necessary moral act.
(6) *Disinterestedness*: Men are expected to achieve their self-interest in work satisfaction and prestige through serving the community interest.	(6) *Interestedness*: Men are expected to achieve their self-interest in work satisfaction and prestige through serving their special community of interest.
(7) *Impartiality*: A scientist concerns himself only with the production of new knowledge and not with the consequences of its use.	(7) *Partiality*: A scientist must concern himself as much with the consequences of his discoveries as with their production; to do any less is to make the scientist into an immoral agent who has no concern for the moral consequences of his activities.[11]
(8) *Suspension of judgment*: Scientific statements are made only on the basis of conclusive evidence.	(8) *Exercise of judgment*: Scientific statements are always made in the face of inconclusive evidence; to be a scientist is to exercise expert judgment in the face of incomplete evidence.
(9) *Absence of bias*: The validity of scientific statement depends only on the operations by which evidence for it was obtained, and not upon the person who makes it.	(9) *Presence of bias*: In reality the validity of a scientific statement depends on both the operations by which evidence for it was obtained and upon the person who makes it; the presence of bias forces the scientist to acknowledge the operation of bias and to attempt to control for it.
(10) *Group loyalty*: Production of new knowledge by research is the most important of all activities and is to be supported as such.	(10) *Loyalty to humanity*: Production of new knowledge by research in the general sustenance of man is the most important of all activities and is to be supported as such; elitism is to be frowned on.
(11) *Freedom*: All restraint or control of scientific investigation is to be resisted.	(11) *Management of research*: Science is a scarce national resource and as such it is to be as carefully managed and planned for as any other scarce resource.

Different Research Styles and Types of Scientists

This section and the next deal with the responses to questions 9, 10, and 13 of the first round. This section deals with the more abstract issue of the differences between various types of scientists. As much as possible I avoid reference to specific individuals. The next section is almost exclusively devoted to the detailed perceptions and personalities of individual scientists.

A constant theme that runs throughout the study is that of the differences, often intense and bitter, between various fields of science, and particularly the psychological differences that are associated with various styles of doing research. The most outstanding of the differences between fields is that between the general descriptive field geologist and the mathematical geophysicist. Whether they tended to lean toward one side or the other, most of the respondents felt there were sharp differences between the styles of thinking in these two fields. The field geologist is not only trained but required to think in broad, global, yet concrete terms; the mathematical geophysicist is trained and required to think in precise, analytical, and abstract terms. It is not a question of one being right, the other, wrong. Both types and fields are necessary. Each is incomplete without the other. Each requires or presupposes the results of the other to do its work.

In the field it is not the function of the geologist to conduct precise controlled experiments but to use his expert judgment, his intuition, so that he can get a broad sweep-of-the-land. His job is to collect rock samples that are representative of the region or of those aspects of the region in which he is interested, so that back in his home lab the samples can be analyzed in detail. As a result, the field geologist is not only trained but actually comes to enjoy working on the basis of hunches, and he prefers fitting the pieces of a huge puzzle together rather than analyzing any one of the puzzle parts in great detail. Because he works by judgment, which is implicit and intuitive, rather than by analysis, which is impersonal and analytical, the geologist cannot always justify his selections of samples to the satisfaction of the rigorous demands of the formalist. Little wonder, then, that to the mathematical geophysicist the field geologist often appears as a sloppy, muddle-headed thinker and all around uncouth character who just mucks about without any rhyme or reason.

The mathematical geophysicist requires vastly different qualities of mind and as a result tends to value and have an alternate con-

80

ception of science. The phenomena of geology are in general so complicated that, if the mathematical geophysicist is to make any headway in attacking them, he must greatly simplify his models of the phenomena. Thus, it is not just that the geophysicist has a preference for treating problems in a rigorous and mathematical manner, he must have had the experience of having dealt with a broad range of physical problems and the gift of mind to know how and when to make the right (that is, fruitful) simplifications.

If to the geophysicist the field geologist is sloppy and uncontrolled in his thinking, to the field geologist the geophysicist is too simplistic. In the interests of building simplified models, the geophysicist is always being accused by the geologist of having left the true essence of the geologic phenomena out of his models. Where the geologist always strives to highlight and to preserve the complexity of the phenomena, the geophysicist always strives to reduce it to manageable proportions. This difference in thinking naturally carries over to the moon, where it has tremendous consequences for the exploration and interpretation of lunar phenomena. The geologist is often accused of being too earthbound, of seeing every lunar phenomenon in terms of earth-based geologic[1][2] processes. The geophysicist rightly raises the point that the geologist may not be justified in extrapolating from the earth to the moon. There is no guarantee that the moon has to be like the earth. The geophysicist is often accused of being ignorant of or ignoring basic geologic processes. The geologist rightly argues that the moon may or may not be like the earth, but that does not mean we should go out of our way to avoid all geologic explanations. To the geologist there are some features — for example, volcanism — of the moon which seem to naturally call for a geological explanation.

This is not to say that every geologist is naturally global and intuitive and every geophysicist naturally analytical and precise or that these qualities are confined solely to geologists and geophysicists. If anything, my use of the terms geologist and geophysicist stands as much for general stylistic or attitudinal differences as it does for specific field or disciplinary differences. Overriding the differences between specific disciplines was a stylistic difference that appeared to be much more basic and fundamental.

Very early in the study it became apparent from the comments of the respondents that there was a fundamental difference among them. A significant part of that difference could be captured along a dimension whose essential underlying quality was "willingness to extrapolate beyond the available data." In any social grouping, there are always those who prefer to stay close to the facts and those who prefer to venture beyond them, even at the risk of ignoring them. As one of my friends who was himself a physicist put it in remarking about some of my general findings, "Doing science is like building a house out of playing cards. Some try to see how high they can build the edifice before it topples over; others wouldn't dream of building the second floor before they completely understood the first floor." The scientists who fell toward the speculation end of the spectrum were scientists who enjoy speculation, perhaps even relish it. These scientists were much more prone to make wild intuitive leaps beyond the data. They more readily tended to see the positive advantages of speculation in science and to encourage its development. They certainly tended to speak much more glowingly of it. The scientists who fell toward the hard data end of the spectrum tended to disparage speculation and wild theorizing, which some referred to as "finger-painting-in-the-sky." These scientists more readily tended to see the negative aspects of speculation and to discourage its widespread application in science. They certainly spoke much more glowingly of sticking close to the data, of the importance of getting good numbers in science. Much more was operating, of course, than a simple, isolated attitude toward speculation. The scientists who leaned toward the hard data end of the spectrum were reacting as much to what they considered the outrageous personality attributes associated with some of the more speculative types as they were to the pure, abstract qualities of speculation itself. Here, again, it proved impossible to keep the exploration and assessment of a supposedly abstract quality like speculation completely apart from a discussion of the personalities of specific individuals.

I am not talking about a purely hypothetical dimension when I speak about an underlying "willingness to extrapolate." First of all, there is ample evidence of the differences between the various types in the literature of geology. An extremely interesting article by T.C. Chamberlin, one of the early deans of American geology, is a clear and forceful expression of the many attitudes to which I have been alluding.[13] In his paper, "The Method of Multiple Working Hypotheses" [67] — or, as it was referred to by a number of the respondents who recommended the paper to me, "The Method of Multiple Working Prejudices" — Chamberlin notes that every investigator naturally tends to give one explanation for every complex phenomenon and that explanation then tends to get converted into a favored theory: "Briefly summed up, the evolution is this: a premature explanation passes into a tentative theory, then into an adopted theory, and then into a ruling theory" [67, p. 755]. In order to counteract this tendency, Chamberlin proposed the idea of multiple working hypotheses — the use of the hypotheses to uncover the facts rather than to support any particular hypothesis or any single ruling theory:

> The working hypothesis differs from the ruling theory in that it is used as a means of determining facts, and has for its chief function the suggestion of lines of inquiry; the inquiry being made, not for the sake of the hypotheses, but for the sake of the facts...
>
> As in the earlier days, so still, it is the habit of some to hastily conjure up an explanation for every new phenomenon that presents itself. Interpretation rushes to the forefront as the chief obligation pressing upon the putative wise man. Laudable as the effort at explanation is in itself, it is to be condemned when it runs before a serious inquiry into the phenomenon itself. A dominant disposition to find out what is, should precede and crowd aside the question, commendable at a later stage, "How came this so?" *First, full facts, then interpretations* [67, pp. 754, 755; emphasis added].

From an extreme or pure speculative type, the admonition might well be exactly reversed: *First, speculation and free rein to the imagination, then facts, if ever facts at all.* From one, such as myself, who tends to view things in systems terms [16, 72], the admonition is: *"There is no first! Facts and interpretation are inseparable.* It is as meaningless to talk of facts without some form of interpretation as it is to talk of interpretations unconditioned by at least one fact." N.R. Hanson put it as follows: "Scientific observation and scientific interpretation need neither be joined nor separated. They are never apart, so they need not be joined. They cannot, even in principle, be separated, and it is conceptually idle to make the attempt. Observation and interpretation are related symbiotically so that each conceptually sustains the other, while separation kills both [168, p. 99]."

The second sense in which the "willingness to extrapolate" is not a purely hypothetical dimension is more dramatic. During the course of the interviews and completely on his own, one of the respondents (Scientist Reed) went so far as to verbalize a taxonomy of scientific types that he had formulated in order to organize his perceptions of the different kinds of scientists he had observed over the years. Although the typology was admittedly crude, Scientist Reed suggested that the body of scientists could be fitted into a schema of three types. Type I scientists excel at extrapolating from data. They excel at theorizing; they relish it. Although they are often fine, detailed experimenters and even at times enjoy experimental results or numbers for their own sake, theorizing is obviously their most pleasurable and exalted task. Type I scientists relish the bold intuitive and theoretical leaps always required in extrapolating from incomplete data to a comprehensive and encompassing theory (see the norm *exercise of judgment* on the counter-norm side). Type III scientists represent the other end of the spectrum. Here, numbers are often worshiped entirely for their own sake. There is a preoccupation, even an obsession, with data gathering. There is often an extreme disdain of theorists who deal with highly inferrential and abstract concepts. Speculation or extrapolation from data is valued little and only engaged in when the data clearly warrant such extrapolation, and then only with extreme caution. Type III scientists are often seen as brilliant but extremely narrow and specialized experimenters. Some are regarded as nothing more than "super-technicians with PhDs." (In fairness, it was noted that theorists can often be just as narrow. But it was the consensus of the sample that it was much more difficult for a theorist to be a narrow specialist than it was for an experimenter. Theorizing on something as broad as the origin of the moon requires, by its very nature, that a scientist be familiar with, if not competent in, several diverse scientific fields.) Type II scientists represent a good mix of theory plus experiment. These scientists are as capable of doing competent experimental work as they are of engaging in modest theorizing and extrapolating activities. From time to time, they could even rise to bold feats of theorizing and extrapolation, but in general they represented the middle ground, running to neither of the extremes represented by Types I and III.

Scientist Reed's typology was one of the prime topics to emerge in response to the focused interview questions. It was also one of the prime topics that later became a standardized interview question — a question which was later put to the rest of the sample. There was widespread agreement with the notion that the types captured some of the important differences between scientists, although many of the respondents remarked on the crudeness of the typology. A few even suggested a greater fineness to the typology and expanded it to include more than three types. In order to learn more about the sample (as well as to put the typology to use and to further test), each respondent was asked whether the names of any scientists came quickly to mind as outstanding examples of each of the types. Very few of the respondents were unable to respond quickly with at least one name for each of the categories. The responses to these questions are discussed in more detail in the next section, where I go into the perceptions of specific individuals. Below are the comments of some of the scientists.

Scientist Quay on the virtues and importance of speculation:

So many people make the mistake of thinking what they know about the earth to be true everywhere. One mustn't think that the things which are the closest to you are the natural state of things. This is the trouble with geologists. They tend to be too earthbound. This is why I've wanted the planning committees to be broader in composition. They tend to be dominated by too much of one kind of thinking.

It seems to me that my strength in science lies in tackling problems that involve a wide range of fundamental physical processes. I don't like to do anything that involves a very high degree of specialization.

Speculation is part of the scientific process. It's just as serious a scientific subject as any other. The reason why speculation is apt to be poorly regarded is that it is apt to be confused with just bad inexact science. I would be very disturbed if anyone ever tripped me up on having done a wrong piece of physics in one of my speculations. I pride myself that whatever physical calculations are involved in my speculative thinking are right.

Speculation is practically the reason for my success. I hope this is what I instill, wide thinking on a subject. And in fact this is quite easily done. One just has to raise a few appropriate questions to remedy the flat way of thinking. Science students have to realize that there's an entirely different way of looking at something that they haven't even begun to think about.

Scientist Reed on the different types of scientists:

Although it's my idea, others have agreed with me that there are basically three types of scientists: (1) people who collect data and publish it with little interpretation [Type III]; (2) people who collect data, publish it, state its significance, and even go a bit further and say you can extrapolate to some degree but who still have their feet firmly on the ground with data as their base, and even in the end point out that there is a lot of it that may be rubbish [Type II]; (3) people who take other people's work, put it together, synthesize it, and come out with a big picture [Type I]. This last type isn't too concerned if they're wrong on this or that, or even if they're totally wrong; they just enjoy playing the game. The third game [Type I] is something amateur scientists often do, people who have revelations, like the Velikovsky idea, not that Velikovsky is stupid; he's very bright, but amateurs like to do that sort of thing. To do the third thing you've got to be pretty damn bright and a particular sort of person. Few do this and those few are pretty good. A lot of people look down on them and scorn them. There's quite a bit of derision around the place. I think the first class [Type III] looks down on the third class [Type I] but not so much the other way around.

To Smith a lot of my work is sloppy because I don't do complete experiments to his satisfaction. To Smith it is a much more interesting thing to get something down to the level of 10^{-7} precision than it is to get a rough and dirty test of an idea. I couldn't care less about precision for precision's sake unless it showed me something important in earth science. Smith is a good example of a [Type III]. He is a very careful, good, thorough worker who prides himself on not speculating at all... Getting a PhD almost guarantees you are a [Type II], if not a [Type III], when you are doing your PhD thesis. If a guy wants to blossom out and use his imagination to become a [Type I], he does it very soon after that. If one studies under a [Type I] guy, he is definitely more likely to be a [Type II] himself. I got a lot of my training at organizations A and B, and speculation is very dangerous at these places. At least it's implied. But I grew up essentially in a [Type II] environment...

Regarding speculation, immense extrapolations must be made if a person is to go to the core, the nature of a problem.

Scientist Sloan on types of scientists:

I would further subdivide the classification of scientists into 5 or 6 types. *First* of all, there's the pure data collector. A lot of people who do routine analyses are of this type. [Sloan then named a few names]. There are no truly imaginative geochemists in this class. The *second* class is composed of the kinds of data collectors who interpret data to some extent. These are very standard types in science. Many of them teach at universities and run a research program on the side which goes on for 2 or 3 years. They produce the standard sorts of papers although many a good scientific contribution comes out of them. There are hundreds of these types; they are the standard inhabitants of scientific departments; they do most of the teaching. Their data is very useful. The *third* class is composed of people who are a little more imaginative. They have a shorter period of data collection. They have more than just routine problems to occupy their time. There are hundreds of these as well. [Sloan then mentions several well-known lunar scientists.] The *fourth* class of people still collect a moderate amount of data but only to test very specific ideas, and if one of their ideas works, they very quickly put out a paper on it. They look for interesting problems they can solve with the data they've collected. [Sloan then mentions a number of well-known lunar scientists.] The *fifth* class of people are really the highly imaginative ones doing first-rate thinking on important problems and collecting data on them. They are making the real major advances [Sloan mentions several well-known lunar scientists.] In the last and *sixth* class are the really top-ranking people. [Sloan mentions very few scientists.]

Scientist Tate:

Science in a way is a sort of religion as it is asking the ultimate questions. Apollo is wonderful but it is not philosophical enough. To some people science is a game like Hesse's *Magister Ludi*, which is a caricature of the academic world, science without the philosophical aspects.

Scientist Upsher was almost the complete opposite of Scientist Tate above:

I wouldn't have gone into my field 50 or 100 years ago because it was too unsettled; the problems were not clear-cut. There were no clear ways of settling them. The questions were like the questions of the moon's origin.

The scientists in the lunar business are putting together scrambled, half-baked ideas in a very qualitative way without a rigorous analysis of the consequences. None of the scientists I've talked to would I consider first-rate scientists. There is no equivalent of an Einstein or an Oppenheimer in geology. I can't think of any geologist who is known that really stands out.

Scientist Victor on the difference between theoreticians and experimentalists:

It's very hard to prove something. The theoretician starts off from the point of view that this is so and let's see if you can absolutely disprove it. The experimentalist just says he doesn't know what is so and is going to try and prove something. The position of the theoretician if he's smart, and those are the only ones that survive, is extremely easy compared to the experimentalist who is trying to prove something.

The only way you can have the chutzpah to develop a theory is to have a tremendous amount of self-confidence. What this means is that someone who is an experimentalist or sitting on top of the data is prone to endow the theoretician with scientific dishonesty. Because in order to be a theoretician one has to latch on to those data which agree with his theory and either ignore or explain away the data which disagree with it. One sees this over and over again. At a meeting over in building Y, Jones, Smith, Young and I were present. Out of the three of us who are sitting on top of the data and making the measurements, Young [the theoretician] is doing most of the interpretation, taking the football and running to the goal line. Young seized on one piece of our data [and] says this is proof that the moon was such and such at one time. We find this hard to disagree with; at least it's hard to prove Young is wrong, but we always have the nagging, gnawing doubt that it's based on an unfounded assumption. There's so much circular reasoning in these theories.

NASA has gone to much trouble to get the best analytical people around. By and large these are the people who are more interested in techniques than in building theories. They hold off building theories until they have a maximum of data. With other kinds of phenomena, we and others have had to make a lot of theories on a minimal amount of data. So now suddenly you have the confluence of these two drastically opposed points of view. They are injected into a milieu where the reports of facts are taken like a football and run with by the theoreticians, and the analytical types are absolutely shocked and amazed.

Scientist Xilas:

Just unsophisticated horses-asses disparage the guys who make minute observations or the guys like Jones and Smith who propose broad ideas. You need both types in science.

Scientist Zimmer:

You only engage in speculation when the data aren't clear-cut.

Scientist Baker:

This science is somewhat different from other sciences. In scientific field X, if I have an idea I do some experiments in the lab and check it out in six months or so whether I'm right or wrong. If I'm wrong, I go on and in 10 years or so I build up a good reputation. In this field you don't get confirmation of your ideas until 10 or 20 years go by and then you have such an enormous investment in your idea that it hurts dreadfully to be wrong. That's why it's more subjective in this field. Of course I hope I'm right. I have a heavy investment in my ideas. But it doesn't bother me as much as it would bother others. I sometimes wonder though why I even bothered with this subject.

Scientist Case:

Geology is really a second-rate science. These guys [the earth scientists in the Apollo program] could never hold their own in another science. They'd be crushed by the competition. They've been thrown into the limelight by the Apollo program and they've muffed it. If it wasn't for the Apollo program, you never would have had any reason to hear of these clowns.

Scientist Davis:

The idea that geology is a second-rate science is nonsense. It's just not true. It's a different science than physics and it demands different skills. It may be true that in the past that some second-rate science students have gone into geology, but I believe that the best in geology could hold their own against the best in any science. NASA has taken special care to assemble some of the best scientists in the world in this program.

The preceding comments give a very slight flavor of the strong differences that were manifested. Thus, for Scientist Quay, speculation is an essential element in his basic approach to science,

88

while for another (Zimmer) it is something that one turns to only when the data are not decisive and clear-cut enough to settle the problem. For some scientists (Reed, Sloan) there is a set of different scientific types and there is also a hierarchy among them, while for another Scientist (Xilas), all the types are of equal importance. For one scientist (Tate), a topic is not interesting if it is not philosophical or speculative enough, while for another (Upsher), if a topic is too speculative or unsettled, it is to be avoided ("not gone into"). For one scientist (see Scientist Logan earlier in the chapter), the field of earth science is no worse (that is, subjective) than other sciences, while for another (Baker) it is worse and he wonders why he ever bothered with it.

If anything, the differences between scientists were even stronger than the comments indicate. Throughout the entire study, for every attitude and opinion on the one side of an issue, there was at least one equally strong attitude and opinion on the opposing side. Agreement or consensus may be the distinguishing mark of science for some [60, 486], and it may be the normal state of affairs, but it is not true that science is totally free from conflict or that it should be. For some kinds of scientific subjects or for some kinds of scientific communities, conflict may be more the rule than the exception, and, as it will be argued later, this may be a better operating rule for science than consensus. One of the reasons the role of agreement may have been overemphasized in science is that we may have systematically overlooked precisely those aspects that could challenge it. It is not that scientists naturally agree or disagree but that they do both. By concentrating on one aspect to the exclusion of the other, we may give the false impression that scientists do not do both. If this book tends to concentrate on the disagreement aspects, it is because I believe these aspects have not received their due in the standard treatises on scientific method. When our eyes are dazzled by the shiny end-products of science over which there is likely to be agreement, it is easy to overlook the disagreement happening in the messy process of production.

Because the differences among the types of scientists were so sharp, and even more important, because the idea of different types had been proposed by the scientists themselves, the decision was made to pursue more systematically on future interview rounds the psychological differences among the scientists. Those who are familiar with Jungian psychology will recognize that there is more than just a surface similarity between these types and the generic psychological types of Jung. Thus, the mathematical geophysicist in this discussion is representative of Jung's thinking type; the field geologist, and particularly the speculative type, is representative of Jung's intuitive type; and the hard-nosed experimentalist is representative of Jung's sensation type. In Chapter Five I demonstrate how the results of this round were used to construct an explicit Jungian typology of different scientific types. The typology was presented to the scientists for their self-evaluation as well as for the evaluation of their peers. The results of this section illustrate how the results of the first round influenced the course of future interview rounds.

Nature and Function of Specific Scientists

In going through the comments of the respondents with regard to their feelings toward one another, the extreme volatility of the comments is most striking. For sheer intensity of emotion, vituperation, and overall vindictiveness, none of the other areas under investigation even begin to approach the areas concerned with the scientists' feelings about one another. So many of the remarks and judgments contained profanity and bordered on slander (if they did not cross over into it) that it would be difficult just to count them. Although so many of the comments were extreme, perhaps the most extreme was uttered by one of the more prestigious scientists who, after the discussion had gone on for some time blurted out, "Look *they're* all fucking ass-holes!" The "they" not only referred to some of the key individual scientists in the program but to the general body of scientists. Although perhaps a bit more striking than the average statement, it is important to emphasize that even this extreme statement was not an isolated expression of sentiment.

What is true of the reactions of the sample toward one another in general is even truer of their reactions toward certain key groups of individuals in the sample. The sample's reactions toward each other were strong. Their reactions towards those designated as key individuals were even stronger. For this reason, it became even more imperative to take special precautions to protect their identities. Thus, in discussing the perceptions and characteristics of some of these individuals, I treat them as a body and not as separate individuals. To isolate and assemble in one place the perceptions of some of these individuals would almost be enough to identify them. To discuss them in blocks or in classes of different

90

Table 6.

Contingency Table of the Distribution of the Types of Scientists Against Their Status

	Gross types						
	Speculative/ theoretical I		Combined/ in-between II		Experimentalist III		
	Refined types						
	6	4	5	3	2	1	Totals
Very high status	2	1	5	0	0	0	8
High status	1	2	5	2	2	0	12
Moderate status	0	2	1	7	3	0	13
Low status	0	0	0	2	0	7	9
Totals	3	5	11	11	5	7	42

Note: Contingency coefficient, $C \sim 0.7$; chi-square ~ 45, df = 15, highly significant.

types will be enough to raise doubts about the identity of the subjects. Starting with the next chapter, where the scaled perceptions of some of these key individuals will be formalized, they will be labeled as S_1, S_2, etc. The scales perceptions of the individuals are abstract enough that by themselves they will not be enough to identify who S_1 is.

Table 6 gives the breakdown of the numbers of scientists in each of the different categories. The gross types refer to the 3 types (I, II, III) proposed by Scientist Reed. The refined types essentially correspond to the 6 types proposed by Scientist Sloan.

In the category type 6, we find not only the most pure of the theorists but also those scientists who were almost universally regarded as the most speculative and wildest of the theorists, the ones who created the biggest and the broadest pictures. This quality of intense speculation and wild extrapolation beyond the data was regarded as their most distinctive characteristic. It should come as no surprise that this group was also almost universally regarded as the ones who were most committed to their theories or points of view and, as a result, regarded as least likely to shift. Because of this fact and because this group of scientists was judged more intensively and derisively than any other group in the entire sample, the perceptions of all three of these individuals were systematically explored over each of the remaining rounds. This group of scientists is admittedly *not* representative of the class of theorists in general and even less representative of the general body of scientists. This alone was enough to arouse the ire of a number of the respondents when these three individuals were used later for purposes of comparison and measurement. But then they were used precisely because they were not representative and hence provided an insight into the kinds of scientists who are likely to become most intensely committed to their ideas.

The scientists in class 4 under type I were also almost universally acknowledged as theorists; a few of them were regarded as being of the highest rank. Their distinguishing mark was that they were regarded as being far more controlled than the first group in their theorizing, "more tuned to the data" and less prone to engage in an "orgy of wild extrapolation." Although a few of these scientists were perceived as highly imaginative, even wildly so at times, in general they were perceived as less imaginative than the first class of type II (5), and so I have ranked them somewhat lower.

In the first class of type II scientists (5) we find the great body of outstanding lunar scientists. Here we find men who are capable of doing outstanding experimental and theoretical work. These are the kinds of scientists who are not only able to design and to execute clever and insightful experiments but who are also able to use and to build the theoretical models that are necessary to give their experimental data the significant interpretation it deserves. One scientist out of group 5 was also chosen for systematic exploration over some of the later rounds. Unlike the first three type I scientists, who were not perceived as representative of the class of theorists, this scientist was the one whose name was mentioned most often as a good, representative example of an outstanding type II scientist.

In group 3 we also find a number of outstanding scientists — indeed there are outstanding scientists scattered throughout the sample. In general these scientists were much less highly regarded by their peers for their ability to combine both outstanding experimental and theoretical work than those in group 5. One of the main distinguishing marks of both groups of type II scientists is that within and across these groups we find some of the strongest and most influential professional cliques.

In group 2 we find those scientists who were generally regarded as good but limited experimentalists. These were the men who often did important and significant experimental work but who were perceived as ranging from "somewhat" to "extremely" limited in drawing the full significance of their work. In this group are found the men who are reluctant to extrapolate beyond the data, men who revere "the facts" at hand. They are perceived as lacking either a deep theoretical background or a deep interest in theoretical work. One of these scientists was selected for systematic exploration over some of the later rounds. Time and time again the judgment was made of this scientist, "X is probably representative of the vast majority of scientists in the program;" "X is typical of the average scientist." Through a systematic exploration of the sample's perceptions of X's attributes, I hoped to get a handle on a most difficult area, namely, an assessment of what the *average* scientist is like. Scientist X was most frequently singled out as a typical type I scientist.

Finally, the last category contains those scientists who were most frequently perceived as competent but unimaginative "number collectors." With a few, although important, exceptions, this category also contains those scientists who were referred to least frequently. The scientists in category 6 were referred to *at least once* in some capacity by *every* member of the sample. The scientists in category 1 were virtually ignored. The scientists in category 6 were referred to approximately 30 times more frequently than those in category 1. The high number of referrals to group 6 and group 5 is not surprising, given the results of Table 6. Even with a rough measure of status (or perhaps because of it), there is a relatively high correlation between the status of a scientist and his type.[14] As a general rule, the higher the status of a scientist, the higher his or her perceived theoretical or speculative stance. This is not a causal argument. A scientist's status does not rise just because he is speculative. Just the opposite may be true. It may be that those scientists who are speculative and also achieve high status do so because they are excellent enough to survive the critical onslaughts of their less speculative colleagues. A less speculative scientist would be crushed by the opposition.

In the remainder of this section I discuss the perceptions, positive as well as negative, of the few key individuals we follow throughout the remainder of this book. The perceptions of three type I scientists in category 6 who were perceived as most committed to their favorite hypotheses or points of view, and as a result perceived as least likely to shift or to give up their hypotheses, will be noted first. The positive as well as the negative perceptions are listed side by side to emphasize the conflicting perceptions of these scientists. Again, the fictitious names bear no relation whatsoever to the actual names of specific individuals, and the same individual may be referred to by more than one name.

Negative perceptions about scientists in category 6

Positive perceptions about scientists in category 6

"They have no humility."

"Their papers are public relations jobs. They have an insatiable need for glory."

"They are examples of the lunatic fringe. They're classic examples of the *idée fixe*."

"Case has a screw loose; he's so committed to the idea of P that no amount of data from the moon to the contrary will ever convince him that P is a false hypothesis."

"They have absolutely first rate, fertile minds; they are extremely bright and quick."

"Adams and Baker make people extremely mad but they also spur them on. They are on the creative vanguard."

"Case is completely wrong, often correct, courageous, extreme, original, strictly a theorist, fertile, strong-willed, versatile, strongly and emotionally committed to his ideas. His theories at first seem wild, then reasonable, and always act as a devil's advocate."

"Edwards relishes the spectacular and has a craving for power."

"It is a characteristic of Edwards that he is able to assemble the facts very quickly and then instantly to propose a brilliant explanatory hypothesis."

"Gage is an out and out crackpot; I no longer regard him as a scientist. He does more harm to science than he does good. Science would be better off without him."

"Gage is the scientist's scientist. It is valuable for science to have people like Gage. The system would be the worse off without him."

"Meade is more interested in the imaginativeness of his ideas than their truth." [This judgment was repeatedly made of Meade].

"Meade is interested in both the imaginativeness of his hypotheses and their truth, not just one or the other."

"These guys have cost NASA and the space program untold

"These types have so many ideas that they are not bothered

amounts of dollars because of their wacky ideas."

"Victor has a curious if not perverted pattern of reasoning that goes something as follows: 'Hypothesis: If the moon were P, then Q would be true. Premise: I WANT Q to be true. Conclusion: Therefore P IS true.'" [This perception, which was first offered by one of the respondents, was then put to the rest of the sample for their reaction; in general the disagreements ran about 2 to 1 over the agreements with this characterization; it was felt to be too severe.]

"Adams tried to put words in the astronauts' mouths; he tried to get them to see what he wanted to find."

Negative perceptions about scientists chosen as representative of category 5

"In a polemic, right or wrong, Baker will knock you over; he should have been a trial lawyer."

"Davis measures things to prove other people wrong. Sometimes this gets very unpleasant. He rubs the noses of others in their own mistakes."

if one or two of them turn out to be wrong."

"Victor has done science a service by pushing a line for all it's worth. He couldn't have pursued it as effectively if he were not so emotionally committed to it."

"The weakness of Victor is not that he hasn't changed his mind but that he's changed it too often. This is his weakness. He's too flexible."

"Adams is an energetic, delightful, unorthodox, versatile thinker."

Positive perceptions about scientists chosen as representative of category 5

"Highly respected; very fair and liberal."

"I think Davis believes that a scientist is right in being prejudiced, that it is up to someone else to point out the weakness in a scientist's arguments. The scientist is an advocate although most aren't frank about admitting this like Davis is. Most scientists think they're impartial when they really aren't. One of the best examples of a good theorist and a good experimentalist. He bridges the gap between theory and experiment beautifully."

Negative perceptions about scientists chosen as representative of category 2	*Positive perceptions about scientists chosen as representative of category 2*
"Adams gave a lousy paper at Houston; terrible delivery."	"Adams did a good job; gave a nice paper at Houston."
"Capable but not brilliant."	"A good worker."
"Not that good technically."	"A top PI."
"Case is definitely a type III but a poor one."	"Case is a very good type III. He's a fantastic observer."
"Rigid, dogmatic, narrow, conservative; thinks he's god on earth."	"Not narrow at all. He's modest."
"Steals the work of others and publishes it."	"He's honest as can be."
"Gage never interprets his results on a grander scale. He never goes beyond what he can see."	"Gage has directed a whole new area."

Some general comments on the comments: Type IIs received the most positive comments of any group. In addition, the perceptions about type IIs were the most evenly balanced between positive and negative statements. Type Is and type IIIs received the most negative comments. Because the majority of the scientists doing the ratings were type IIs, it is not surprising that they got the best of it.

It is interesting that those three scientists in category 6 who were perceived as most committed to their theories were perceived as intuitive, speculative, brilliant first-rate minds, abrasive, controversial, and as extremely competitive actors who loved the limelight and had an insatiable need for success. However the consequences those perceptions have for our understanding of the nature of science are even more important. Thus, for example, it is important that the public speaking behavior and style of some of these scientists is so outrageous to their peers that a sizable number of them no longer regard them as scientists (recall the comment, "Their papers are public relations jobs."). It is significant that some of their intuitive leaps are so bold, outrageous (but in a different sense), that many of their peers have come to have an extremely disdainful view of the place of intuition and speculation in science (recall the comment, "Meade is more interested in the imaginativeness of his ideas than their truth"). However, it is precisely because of these qualities that another significant body of their peers regards them as first class scientists whose supreme value is their fantastic imaginations which constantly stir things up and force people to think outside of the conventional molds. In other words, one of the most basic judgments that can be made — about what qualities characterize a scientist — is psychologically dependent; it depends on who is perceiving whom and on what that perceiving type values.

But perhaps the most important implication of the comments in this and the previous section is the powerful illustration of the significant role sociological and psychological factors play in the acceptance and validation of scientific ideas. Thus, being in the right or wrong field of science determines whether it will be easier or harder to get through to other scientists. If one is of the right or wrong scientific or psychological type, that means that one's scientific evidence and arguments will either stand a better chance of just being considered or will be rejected out of hand (as evidenced by the numerous comments such as, "I wouldn't believe or trust any argument or evidence of X's.") Instead of being a purely depersonalized and formal process that can be understood independently of all sociological and psychological considerations, the validation of scientific ideas is *in part* sociological and psychological. By this I mean that what one scientist (S_1) of one psychological or scientific type believes or perceives of another scientist (S_2) of another type affects S_1's assessment of S_2's evidence (E) with regard to phenomenon *P*. To restrict the validation and assessment of scientific ideas to S_2's formal procedures for the gathering and testing of evidence with respect to P neglects some of the most important aspects of the whole process, and thus diminishes our knowledge of the individual parts. This is one of the main conclusions of this chapter, and its implications are far reaching. The implications of the results of this chapter need to be examined for a combined philosophical and social psychological theory of scientific objectivity. This is done in the concluding chapters.

Perceptions of Importance of the Study

Two of the most interesting of the remaining first round questions have to do with (1) how the significance of this study was perceived by the respondents, and (2) when the respondents first became aware of, and as a result whether they became disillusioned with, the actual behavior of scientists. The first question was part of the original set of questions; the second emerged as a result of the interviews.

The question of the perceived significance of this study is important for a number of reasons. It gives a measure, no matter how crude it might be, of the motives of the respondents — of why they were so willing to talk about such personal matters. Approximately a fourth of the sample, in one way or another, communicated the feeling: "I'd like to do what you're doing only I couldn't do it. I'd be drummed out of the profession. We all know the behavior of these guys and we talk about it all the time." Throughout the study I had the distinct impression that for many of the scientists the interviews were cathartic. The interview process seemed to allow many of the respondents to get things off their chests, to talk at uninterrupted length and to a nonthreatening, sympathetic listener about things that had been simmering within for years. It became clear that my volatile questions had not raised things that had not already repeatedly crossed their minds. James D. Watson's book *The Double Helix* [458] had apparently made a strong impact on many of them, at least enough so that a number of the respondents brought it up of their own accord. But there were many motives. No single set suffices to explain such a diverse and complex group. Some scientists were just frankly helpful and thought the project was interesting. But even these respondents seemed to have a strong underlying motive. Nearly every scientist in the sample felt strongly that science needs to be demythologized: "The public, scientific as well as general, needs to see that scientists are no better or no worse than any other group. Scientists have been made into supergods for too long." And finally, there were, as might be expected, those few who seemed motivated by the rare opportunity to tell their side of the story, to get their brief share of the spotlight by disclosing something special about the nature of science. This was particularly evident among those who seemed to be on the fringes of the sample.

The responses to the question of when the scientists first became aware of the actual behavior of scientists and whether this was disillusioning fell between two poles. For the vast majority of scientists in the sample, the process of eye-opening was gradual and seemed to occur at about the time in graduate school when they began to work closely with some of their professors and were thus in a position to observe actual behavior firsthand. For this group of respondents, the actual behavior of scientists, even if extremely irritating, came to be accepted as natural. The most frequently heard expression, if not rationalization, in conjunction with this position was the statement "Scientists are only human." For a number of scientists in this group, science has become no more or less than a job, a profession. The ideals of science as traditionally defined have lost whatever meaning they once had. Science is like any other profession. Success is based not just on what a man has done but on where he has done it and who he knows. Toward the other pole were those scientists who had experienced surprise and shock when they left graduate school and discovered that there was a large discrepancy between what the textbooks idealistically and naively said went on in science and what really went on.[15] This group constituted a much smaller minority. Respondents near the first pole deflated the ideal picture ("Scientists are only human"). Those near the second pole held on to the ideal picture by appealing to some abstract notion like the concept of nature — "Men may not be disinterested, but ultimately nature is." Each pole represents a way of coping with the system.

Concluding Remarks

In spite of the extreme nature of so many of the comments in this chapter, it would be a serious mistake for the reader to conclude that this material is in any way meant to detract from the excellent scientific work done in conjunction with the historic Apollo 11 mission. This book is a look behind the scenes at a side of science that we rarely have the opportunity to observe. That these two sides — the public and the traditionally objective side and the hidden, personal, and seemingly subjective side — can coexist is one of the themes of this book. The existence of such extreme emotions does not prevent scientists from doing excellent scientific work and being reputable and scrupulous. As I will argue later, I believe that scientists may need to have such strong emotions and commitments in order to do excellent work. I am no less impressed — indeed I am more impressed — by the achievements of Apollo 11, given this material. In short, I would strongly object to any interpretation of this material which attempts to cast doubt on the excellence of either the scientists or the scientific work done on Apollo 11. We must relinquish the childish notion that science and scientists are or must be either totally objective or totally subjective. This is too naive a view of how any system actually works. The constant interactions and blends between the two should be our goal.

This study is not meant as a gigantic exercise in muckraking. It may be true that sociologists of the survey research type are professional gossip collectors, and it may also be true that psychologists of the personality assessment variety are professional character assassins, but it is not my intent to collect gossip, to assassinate character, or to muckrake. Even if this were an exercise in muckraking, which it is not, the charge would be irrelevant. The charge of muckraking fails to deal with the implications of this material for our understanding of science. Muckraking is an easy charge to hurl, but it is a poor analysis of the material. The material of this chapter will not be easily dismissed by labels. When science and scientists attain the power they have in our society, when our civilization comes more and more to depend on both, and argues for their preeminence, then — whether we are muckraking or not — we had better examine carefully the psychology, the sociology, the philosophy, the emotions, the thoughts, the fantasies, the beliefs, the opinions, and the workings that underlie these vital forces [292].

I also do not wish to be labeled as an advocate of a subjectivist or irrationalist interpretation of science. This is the problem that those who study the outright subjective side of any subject always face. I do not believe that science is a rambling, aimless, irrational activity. I consider myself too much of an objectivist trying to explicate the rules and structure of science. I just do not think those rules can be understood without a fundamental incorporation of the emotional side of science.

Finally, the material of this chapter raises a number of central questions: Why has the crucial place of emotions in science been neglected for so long [292]? Why has the topic not been given the study it so rightly deserves? Why have emotions been perceived in a neutral light at best (thus the disinterested and not the interested observer, thus neutral observations and not interest-directed observations) and most commonly as something to be controlled if not eliminated all together? I shall return repeatedly to these questions, for they are among the preoccupations of this book.

Chapter Four

Ideal Scientist: Apollo 12 and 14

"People are more interesting than rocks. At least I feel that they are. Most of my colleagues in geology probably feel just the opposite, that rocks are more interesting than people."

Quote from one of the scientists in the sample

The previous chapter presented a substantial portion of the major results of the first round of interviews. The next three chapters present some of the major results of the remaining three rounds of interviews. The basic purposes of these remaining rounds were as follows: (1) to clarify some of the many issues and findings of the first round by pursuing them in a much more systematic and quantitative fashion, and (2) to chart the development and change of some of the attitudes associated with these issues by means of repeated measurements over time.

In this chapter an attempt is made to assess and to conceptualize more systematically and quantitatively the various types of scientists that were discussed in the last chapter. This is done by means of a variety of psychological instruments that locate the various types of scientists in relationship to one another in a number of many-dimensional psychological spaces. The axes or dimensions of these spaces are various psychological attributes in terms of which each member of the sample was asked to scale his perceptions of some of the specific scientists who were designated in Chapter Three as representative of the various types of scientists. A variety of statistical functions were then applied to the perceptions of the entire sample, to transform the perceptions of these types into distinct points in a number of psychological spaces. The various configurations of these types in the variety of spaces tell us a number of interesting things about the psychology or definition of these types. The configurations also tell us something about the psychology of the respondents and so about the psychology of scientists in general. The results of this chapter are then compared to some of the existent literature in the psychology and sociology

101

of science for purposes of further analysis and clarification. The results of this chapter both support and cast doubt with respect to a number of previous findings and hypotheses.

In Chapter Five, an assessment is made again of the positions of the respondents with respect to a number of scientific issues pertaining to the properties of the moon. Using numerical scales, one can measure quantitatively where the respondents stood with respect to these issues, before and after Apollo 11. By comparing the before and after positions of each respondent, certain measures can be computed which, in information-theoretic terms, give the average amount of shift of the entire sample with regard to the issues. The next quantitative measurement indicates how those specific scientists who were selected as representative of the various types of scientists in Chapter Three were perceived as falling on specific scientific issues, and how committed they were perceived as being to their positions. This exercise was also repeated on the last round in an attempt to assess the amount of perceived shift that was associated with these few highly select scientists. Finally, the results of an alternate approach to the conceptualization of the various types of scientists is presented. Drawing upon Jungian personality theory, a number of character sketches were constructed of different kinds of scientists. Each respondent was asked to rate himself as well as some of his peers in terms of those sketches. It is interesting and informative to compare the results from different ways of assessing scientific types. And in the enormously difficult area of personality assessment, it is imperative that multiple measurements or assessments be employed.

Finally, some of the results of the last round of interviews are presented in Chapter Six. The themes here are varied and many. For example: (1) the sample's judgment of the most significant results and findings to emerge from the Apollo missions to date; (2) the sample's evaluation of the performance of NASA in planning and managing the Apollo missions; (3) the sample's judgment as to whether any serious errors were committed in the planning of the Apollo missions and the implications, if any, for the planning of future, large-scale, scientific missions; (4) the reactions of the respondents to this study — for instance, whether it made them think about anything concerning science that they had not thought of before; and (5) their reactions to this investigator. It would be a shame to interview some of the most important participants in the Apollo program and not raise issues like (2) and (3) above, which have to do with some of the broader implications of the program and which bear on the general issue of science planning. This is particularly true if, after the completion of Apollo 17, we are not to return to the moon for at least another decade. One would certainly hope that the feelings of some of the most important participants would be considered as a novel and important source for the evaluation of the particular program and for the planning of future programs. It certainly would be parochial to confine a study such as this solely to an examination of the psychology of scientists and not raise other issues.

However, it must also be noted that I have had to confine the treatment to a relatively few issues out of the many that were explored in depth. The above outline or preview of the material in the forthcoming chapters barely begins to exhaust the full range of issues that were raised. For reasons of space and in order to avoid overwhelming the reader, only a subset of the total issues can be dealt with in this book. Thus, for instance, in round III there were a large number of attitudinal items having explicitly to do with the philosophy of science. The question that is of potential interest is whether the responses to these items are sufficient to differentiate among and define different types of scientists and whether the responses are sufficient to establish that different scientists do indeed have different concepts or models of science. As important as the systematic investigation of these issues is, it must be deferred to future reports. In this book, I deal with a highly selective examination of the responses to a very few of these and other items. Although a variety of themes will continue to be pursued throughout the remainder of this book, for the main purpose of lending more coherence to discussion I shall focus in particular on one of the major emergent themes. I shall be primarily concerned with pursuing a more systematic and quantitative definition of various types of scientists and with examining further the implications which these different types have for our conceptualization of science.

Modified Semantic Differential Technique

The semantic differential as first developed by Charles E. Osgood and his colleagues [338] is a technique for measuring the meaning(s) of a concept and comparing it (them) with respect to the meanings of other concepts. In the typical application, a group of subjects is presented with a concept or a number of concepts

103

whose meanings are to be determined. Under each concept are a series of scales bounded at each end by a single adjective. The adjective pairs that comprise each scale are antonyms. The scale between every pair of adjectives typically has seven places on it. The subjects are given the concepts one at a time and instructed to place on each scale a single check mark which best expresses their feeling towards the concept being rated. For example, consider the concept *Father* which is to be judged on the two scales strong/weak and kind/cruel:

Father

strong: — — — — — — — : weak
cruel : — — — — — — — : kind
1 2 3 4 5 6 7

If a subject considered the concept Father more strong than weak, he would check the spaces 1, 2, or 3. A check in the 1 space would mean that a subject considered the concept Father to be at the extreme end of the strong side of the scale. If a subject considered the concept Father neither strong nor weak, or both equally, or the scale irrelevant to the concept, then the subject would check the 4 space. This procedure is repeated for each scale under the concept until there emerges a profile, or a *semantic*, connecting the points that have been checked on each scale. If a number of concepts are rated on the same set of scales, the profiles between concepts can be compared. Further, from the scores on each scale, a generalized Pythagorean distance can be computed between any two concepts. From these distances, and with the use of specific statistical functions, the concepts can be located in relationship to one another in a variety of spaces. The reader who is interested in pursuing the details of the procedure as well as a critical review of the pros and cons of the technique is referred to the literature [413, 416].

This is a very brief description of the essentials of the procedure and the typical way of administering it. The procedure that was actually used in this study was a variant. Throughout the entire rating process, each respondent was encouraged to comment as freely as he wished on what the adjective-pairs that were used meant to him. In effect, each scale of opposing adjective-pairs constituted a standard stimulus to which each subject was asked to respond with both his free comments (which were recorded on tape for later analysis), and his supposedly more objective check marks on a piece of paper. The reason for my adopting this procedure has to do with my critical assessment of Osgood's concepts of *meaning* and of *objectivity*.

For Osgood [337, 338], both the meaning and the objectivity of the semantic differential (SD) are essentially confined to the check marks on the scales (and their "impartial analyses" according to "fixed rules"), which define the profile under a concept.[1] As I will attempt to make clear in Chapter Seven, Osgood's notion of objectivity is very restricted. The objectivity of the SD cannot be reduced solely to check marks on paper or to impersonal technical means for analyzing those marks. Likewise, the meaning that the SD measures cannot be similarly restricted. It is true that Osgood distinguishes between *denotative* and *connotative* meaning [337]. The former (denotative) supposedly deals with the relations between signs (symbols, words) and their referents (the thing to which the signs refer). And the latter (*connotative*) is supposed to deal with the relation between signs and their users. Osgood contends that it is presumably connotative meaning that the SD measures [416, p. 132]. Whether this is true or not, critics, most notably Weinreich, have pointed out that "there is missing the linguistically crucial domain of relations between signs and other signs such as is expressed in a (non-ostensive) definition" [465, p. 139]. Whatever meaning (connotative or not) the SD measures, it is still with respect to the basic meanings of the adjective-pairs. Now, one can take the meanings of the basic adjective-pairs as *primitives*, as basic terms that are the same for all respondents. However, even if this is the case, this still presupposes that we have some knowledge of the primitive meanings of the adjective-pairs that are being considered in the rating process. It would seem important, if not interesting, to elicit at some point in the process of the development of the SD the meanings that the adjective-pairs have for the subjects as they go through a SD exercise.[2] Indeed, I would argue that some of the most interesting data that one could collect are the verbalizations that are produced by sophisticated respondents (such as those in this study) as they go through the procedure.

The respondents were also encouraged, for a number of other reasons, to verbalize their thoughts as they went through the various SDs. They were asked to comment freely on the appropriateness or relevancy of the particular dimensions to them — that is, on the adjective-pairs. As is true for any structured instrument or questionnaire, in making up an SD the investigator has to choose dimensions that he thinks are relevant to the concept being scaled. But what is relevant to the investigator may not be relevant or interesting to the respondents. In other words, the respondents are forced to respond in terms of the investigator's categories of meaning (frame of reference), which may or may not be the same as their own [413]. Instructing the respondents to comment freely is one way of getting at their own categories of meaning. It is also a potentially important way of securing the cooperation and interest of respondents who may not be socialized into the filling out of structured instruments [413]. For these and other reasons, the respondents were encouraged to respond verbally to every instrument that was presented to them throughout the entire study, whether it called for written or numerical responses or not. Such a technique also helps to defuse some of the apprehensiveness that may be connected with the filling out of the more volatile items to be described later.

The first SD concept that was given to the sample for their rating was that of the Ideal Scientist; 45 respondents rated 27 scales, which were a combination of various general personality attributes plus some specific attitudes or attributes that more directly pertain to science. The purpose of this first SD was to try to get at each respondent's general concept of science and scientists without focusing in on any particular individuals. After the respondents had gone through this exercise, six considerably shorter SDs of 10 scales each were given to them. The respondents were asked to rate 5 specific scientists chosen from round I as representative of the various types of scientists. In addition, a sixth SD (Yourself) consisting of the same 10 scales was also given to each respondent to rate himself.

The Ideal Scientist, the names of the five types, and the general heading Yourself constituted the various SD concepts that were rated. The numerical scale profiles plus various statistics computed on those profiles constituted the quantitative aspect of the assessment procedure. The free-flowing verbal responses of the subjects as they went through the SDs (concepts, scales) constituted the qualitative aspects of the assessment procedure. The Ideal Scientist (that is, the concept and the adjective-pair scales) was developed by my friend and colleague Vaughn Blankenship. The Ideal Scientist is part of a much larger national study which Blankenship has been conducting on the political behavior and attitudes of scientists [45]. The profile scores of Blankenship's respondents on the

106

Ideal Scientist SD have been compared with their profile scores on other SD concepts (Ideal Lawyer, Ideal Supreme Court Justice, Ideal President) in order to test various hypotheses about role perceptions. The majority of the attitudinal items at the end of the questionnaire[3] that was administered in round II were also derived from Blankenship's study. Although it is not carried out here, the possibility of comparison exists between Blankenship's sample and the sample in this study. Blankenship's sample represents a much larger study of a wider cross-section of different kinds of scientists; this study represents a more intensive study of a smaller, more tightly knit group. Also, this study differs in that qualitative, verbal responses were obtained from the respondents in addition to their numerical scores. In Blankenship's study, purely quantitative data obtained from questionnaires were used.

Finally, it should be noted that I do not consider myself an ardent advocate or user of the SD. I feel it is merely one out of many social science techniques for measuring meanings and attitudes. Its use is neither to be encouraged in all situations nor to be avoided or rejected out of hand.

Ideal Scientist Profile

Figure 2 shows the semantic profile that results from connecting the means (averages) of the 45 respondent scores for each scale. This profile is represented by the heavy solid line that meanders down the middle of the figure. Also shown is another profile that I formed by translating the concept of the Storybook Scientist of Chapter One into the terms of each scale. This hypothetical profile (from here on referred to as the Hypothetical Stereotypical Scientist, or HSS for short) is represented by the dashed line connecting the circles. One of the main questions of this chapter is whether the scientists' semantic profile is sufficiently far from the HSS profile so that we can say that the scientists are rejecting the Storybook version of science once again and through means that are different from those employed in the last chapter. In terms of statistics alone, the answer to the preceding question is a firm yes. Applying Student's t-test to each of the SD scales shows that, in 26 out of the 27 cases, the difference between the two profiles is great enough to imply rejection of the HSS at a very high level of statistical significance ($0.0025 \gg \alpha$). However, because the term *rejection* has such strong psychological overtones, the question is

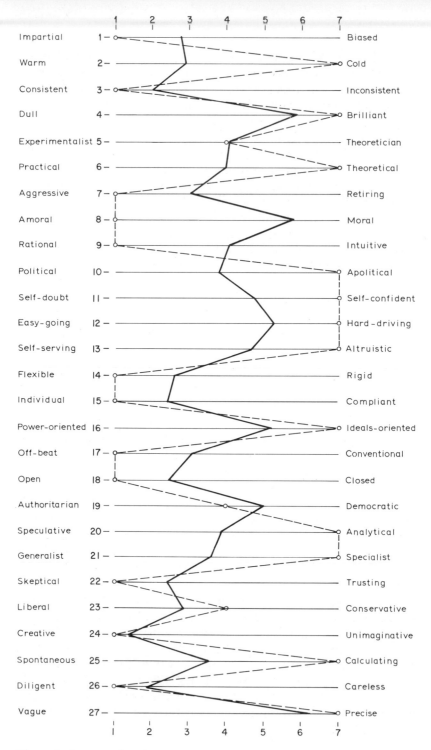

Figure 2. A comparison between the mean values of the sample respondents and a portrait of the stereotypical scientist.

not whether rejection is justified from a statistical point of view, but whether it is also justified from a psychological point of view.

This concern is justified by the fact that the concept of the HSS was not presented directly for the sample's acceptance or rejection, and so we cannot necessarily assume from the statistical rejection of each of the 27 scales taken separately that the concept of the HSS as a whole would be rejected. The rejection or acceptance of the HSS is dependent on the way it has been represented on the SD profile. With regard to the first point, the comments of the last chapter and the verbal comments to the SD scales to be discussed in the next section of this chapter support a rejection of the HSS concept as a whole. From the comments of the respondents it is clear that much more than isolated individual scales were being rated. Nearly every respondent evaluated the concept of the Ideal Scientist with some general concept of science or scientists in mind. A number of the respondents used the abstract idea of the Great Scientist in filling out the scales. A few even had some specific Great Scientist in mind and used this particular case as a guide. The majority of the respondents, however, seemed to use some composite image of their peers. Most often this consisted of a range of possible Real Scientist types. A number alluded to the notion of the Mythical or Fictional Scientist and indicated by their comments that they were filling out each scale in terms of their acceptance or rejection of the properties of this mythical creature. For example, one scientist noted: "According to the Mythical Account of Scientists, one ought to put a check mark here on this scale [either a 1 or a 7] but this is not the way scientists actually are and there are good reasons for not having scientists, even 'ideal' ones, behave like this; so I'm going to put a check mark here [somewhere in the 2 to 6 range]."

There is admittedly more than one way of representing the concept of the HSS on the 27 scales. Any representation is complicated by the fact that the adjective-pairs have a multiplicity of meanings depending on such factors as the context in which they are taken — for example, whether they are to be applied to the Ideal Scientist's personality when he is doing science in the laboratory, or when he is teaching, interacting with his colleagues, at home with his family and friends — or if they are to be taken across the board over all contexts. The concept of the HSS is plagued by various subtleties. Finally, it is not clear whether the concept of the HSS applies to all of the 27 dimensions and thus can be represented in terms of all the scales. Nevertheless, this

Table 7.

Scale Profiles of the HSS and HAS

Ideal Scientist Scale number	Hypothetical Stereotypical Scientist — HSS		
	Scale scores	Corresponding adjective	Basis for assignment of scores, adjectives
1	1	Impartial	CN
2	7	Cold	Masculine, CN Image
3	1	Consistent	Masculine, Image
4	7	Brilliant	Image
5	4	Both exper.-thrt.	Insufficient reason
6	7	Theoretical	Image
7	1	Aggressive	Masculine
8	1	Amoral	Image, CN
9	1	Rational	Image, masculine
10	7	Apolitical	Image
11	7	Self-confident	Image, masculine
12	7	Hard-driving	Masculine
13	7	Altruistic	CN
14	1	Flexible	CN, Image
15	1	Individualistic	Image, CN
16	7	Ideals-oriented	CN, Image
17	1	Offbeat	Image, CN
18	1	Open	Image, CN
19	4 (?)	Both? authoritarian—democratic	Conflicting meaning
20	7	Analytical	Image, masculine
21	7	Specialist	Image
22	1	Skeptical	Masculine
23	4 (?)	Both? liberal-conservative	Conflicting meaning
24	1	Creative	Image
25	7	Calculating	Masculine
26	1	Diligent	Image
27	7	Precise	Masculine, image

CN = One or more of the conventional norms on the rational side (see Chapter Three).

ON = One or more of the opposing norms on the irrational side (see Chapter Three).

Image = Derived by implication from previous studies on the perceived characteristics or image of scientists.

Contrary = Value represents the most extreme opposing HAS score to the HSS assigned score.

110

Hypothetical Anti-Stereotypical Scientist — HAS

Scale scores	Corresponding adjective	Basis for assignment of scores, adjectives
7	Biased	ON
1	Warm	Feminine, ON
7	Inconsistent	Contrary
1	Dull	Contrary
4	Both exper.-thrt.	Insufficient reason
1	Practical	Contrary
7	Retiring	Feminine
7	Moral	Contrary, ON
7	Intuitive	Feminine
1	Political	Contrary
1	Self-doubt	Contrary, feminine
1	Easy-going	Feminine
1	Self-serving	ON
7	Rigid	ON, Contrary
7	Compliant	Contrary, ON
1	Power-oriented	ON, Contrary
7	Conventional	Contrary, ON
7	Closed	Contrary, ON
4 (?)	Both? authoritarian-democratic	Conflicting meanings
1	Speculative	Contrary, feminine
1	Generalist	Contrary
7	Trusting	Feminine
4 (?)	Both? liberal-conservative	Conflicting meanings
7	Unimaginative	Contrary
1	Spontaneous	Feminine
7	Careless	Contrary
1	Vague	Feminine, contrary

Masculine = Masculine *stereotypical* value or characteristic.
Feminine = Feminine *stereotypical* value or characteristic.
Insufficient reason = Insufficient reason to differentiate between either end of the scale.
Conflicting meanings = This author unable to decide between which of the many meanings should be assigned — uncomfortable with either end of the scale.

does not mean that the representation adopted here is arbitrary or was constructed without any guidelines. On the contrary, a number of guidelines were followed (see Table 7): (1) Whatever concept of the HSS is represented should be consistent with the respondents' concept of the HSS as indicated by their comments in the next section. (2) As much as possible, the concept of HSS ought to follow from or at the very least reflect the majority of the conventional norms of science discussed in Chapters One and Three. (3) The concept of HSS ought to take into consideration previous research on how various groups actually perceive scientists [34, 303]. And (4) when in doubt with respect to the location to be assigned to an HSS point on a scale, the point ought to be assigned in such a way that it reflects the fact that science embodies a predominantly masculine orientation toward the world [70, 285]. To a far greater extent than we realize, science represents the masculine spirit transformed into systematic method and dressed up in formal, depersonalized rules. It is thus no accident that the terms which are used to describe the HSS are, more often than not, masculine terms. For example, we use such terms as *cold* (this also follows from the notion of the impersonal, detached, aloof observer), *hard-driving, aggressive, skeptical* (as opposed to *trusting*, which is a feminine attribute in a traditional or conventional sense of femininity), *calculating*, and so on. More will be said about this later in this chapter as well as in Chapter Six.

The HSS profile ought to be consistent with the sample's concept of the HSS as revealed by their comments. However, it should be noted that, although the sample rejected the concept of the HSS, they still took it into strong consideration in framing their responses. And, in fact, a glance at Figure 2 shows that the respondent profile is closer to the HSS profile than to the contrary concept which would be represented by the collection of points antithetical to the HSS profile points. The sample strongly rejected the concept represented by the HSS profile. They rejected even more strongly the contrary concept of, for want of a better name, the Hypothetical Anti-Stereotypical Scientist or HAS (see Table 7). And so one is tempted to say that the respondent profile shows a greater acceptance of HSS, and by inference, an acceptance of the conventional norms of science of Chapter Three, and a greater rejection of HAS, and by inference, a rejection of the counter-norms of science of Chapter Three. This is true if the sole criterion of acceptance or rejection is one of profile closeness, and if it is one set of norms over the other. The decision to accept or to reject any hypothetical or abstract profile cannot be made purely on the basis of any formal or statistical test. Indeed, the sample profile is far enough from both ends of each scale in the statistical sense to reject any one of the 2^{27} possible profiles that could be formed by connecting the extreme ends of the scales at random. To choose either the HSS or the HAS profile to the exclusion of the other would be to ignore the verbal comments of the respondents. Those comments indicate that, in placing a check mark on each of the scales, the respondents seriously considered both ends of the scale. The respondents saw the advantages to science of considering both ends of the scales as norms, and conversely they saw the dangers that would accrue to science by following either exclusively.

Verbal Responses to Scales

Scale 1: Impartial—Biased

Of all the scales, the verbal responses to the impartial—biased dimension show the greatest attitudinal rejection of the HSS concept. Many of the responses parallel the comments of the last chapter where the positive function of even extreme commitment in science was stressed. However, it is clear from the comments that complete impartiality was recognized as the commonly perceived norm of scientific life. It is also clear that it was perceived as an idealized, if not mythical and naive, norm. In this sense, complete impartiality is rejected both as an actual fact of scientific behavior and as a desirable end or ideal. However, general rejection of the HSS or impartial position, as represented by a score of 1 on the scale, does not imply an unqualified endorsement of the converse position, the HAS, as represented by a score of 7. The responses indicate a complex tug-of-war between two opposing sets of norms:

Scientist Adams: The concept of the impartial observer is an absurdity.

Scientist Baker: One has to develop a position, take a stand on it [as one would] in a court of law, and then see whether it's defensible or not. To be forever totally impartial is not possible. One should at least temporarily defend a position, if only to reduce the number of ultimate possibilities that any one person has to contend with in his own mind.

113

Scientist Case: The Ideal Scientist should obviously be on the impartial side of it [the scale]. However, real scientific contributions are very commonly made by a single-minded, highly biased individual with a devil-may-care, the hell-with-them-all attitude.

Scientist Davis: It's all right for a scientist to be rather strongly biased while he's pursuing an idea; he should not be indifferent to the various alternatives he's trying to decide between, but he has to be objective enough to discard an alternative that runs into difficulty. This means he has to be able to switch back and forth between biased and impartial. I'd like to be able to check both ends of this thing, say 2 and 6, simultaneously. Within the constraints of this questionnaire, I'd check, 3 but it's not as black and white as this thing makes it appear.

Scale 2: Warm—Cold. Some representative responses:

Scientist Gage: I take this [scale] as having the personality to get along with all kinds. In this case, you can have the whole range. All the scale points are possible. Some who are cold tend to be extremely impartial, whereas some who are quite biased can be very warm.

Scientist Hall: This scale is completely irrelevant to science. I'd prefer to leave it blank. I wouldn't check any of the spaces.

Scientist Ingram: I couldn't give a damn about this; it's not interesting.

Scientist Jones: This dimension is irrelevant in the sense you can have all kinds of types. Personally I like warm people more than cold people but both can be good scientists.

Scientist King: An ideal *person* would be warmer than colder. You're really talking about something here that applies to people not to scientists in particular.

Scientist Logan: This has nothing to do with science.

Scientist Meade: A scientist who is dispassionate and cold, allowing his whole life to be run this way, is going to lack the motivation to do science for the reason why science is done, for the benefit of the human race. I'd tend to lean toward a warm scientist if you can get such a person; he can run into problems if he's too warm. I'd put it in the middle [checks 4].

Scientist Nolan: I'd check 2 or 3. If a scientist is dealing with coworkers, he should be warm. If he's dealing with a subject, he should be cold; you cannot be involved emotionally in the cold, scientific interpretation of facts.

Scientist Oaks: A person should be warm, but not so warm that he can't protect himself.

114

Table 8

Adjective-Pairs by Groups

Group 1 General personality variables	Group 2 Scientific personality variables	Group 3 Cognitive style variables
(2) warm—cold	(1) impartial—biased	(5) experimentalist— theoretician
(4) dull—brilliant	(3) consistent— inconsistent	(6) practical— theoretical
(7) aggressive—retiring	(14) flexible—rigid	(9) rational—intuitive
(8) amoral—moral	(18) open—closed	(20) speculative— analytical
(10) political— apolitical	(22) skeptical—trusting	(21) generalist— specialist
(11) self-doubt— self-confident	(24) creative— unimaginative	
(12) easy-going— hard-driving	(25) spontaneous— calculating	
(13) self-serving— altruistic	(26) diligent—careless	
(15) individualistic— compliant	(27) vague—precise	
(16) power-oriented— ideals-oriented		
(17) offbeat— conventional		
(19) authoritarian— democratic		
(23) liberal— conservative		

The responses to this scale are most interesting, and even predictable to a point. A glance at Figure 2 (see also Table 8) shows that the adjective-pairs break down into three rather distinct groups, even though there is a considerable degree of overlap among them: (1) general personality variables or dimensions, such as warm-cold and aggressive-retiring, which apply to people in general irrespective of their professional roles and are not confined specifically to the concept of scientist; (2) scientific personality or attitudinal variables, such as impartial-biased, consistent-inconsistent, vague-precise, which relate more specifically to the role-concept of scientist; and (3) general congnitive style variables

115

(styles of reasoning), such as experimentalist-theoretician, practical-theoretical, rational-intuitive, speculative-analytical, and generalist-specialist, which relate even more directly to the role-concept of scientist. The overlap between groups (2) and (3) is particularly great. Many of the dimensions in group (2), such as flexible-rigid, open-closed, creative-unimaginative, diligent-careless, vague-precise, could just as easily be shoved into group (3), because they also pertain to styles or modes of reasoning. The difference is that the pairs in group (2) are more pejorative and value-laden than those in group (3), and, conversely, those in group (3) are more abstract and cognitive than those in group (2). (See Table 8 for a complete listing of the adjectives in each group.)

Because the adjective-pairs can be sorted into various groups, it is not surprising to find that a sizable proportion of the sample regarded the warm-cold dimension as not especially applicable to or even uniquely characteristic of their concept of the Ideal Scientist. What is surprising is that the dimension was judged irrelevant. To say of a particular dimension that it is not uniquely applicable or that it is independent of other dimensions is one thing. To voice the stronger judgment that it is irrelevant or not interesting at all is quite another, particularly when that dimension is unequivocally one of the most important dimensions in terms of which we judge and assess our fellow man [26]. This reflection becomes all the more striking as soon as we point out that, of all the scales, the warm—cold dimension was perceived as the most irrelevant. And in fact scale 2 elicited the largest number of blank or omitted responses of any scale. A full 31 percent of the sample explicitly voiced the judgment that they considered the scale irrelevant enough to their concept of the Ideal Scientist to leave it blank. A blank response was used by these respondents as a stronger indicator of the irrelevancy of the dimension than a check in the 4 space would have expressed. In this regard, I should remark that there were a number of scoring idiosyncracies in the patterns of the respondents that are not normally found with the use of a SD. One of the prime reasons is that such idiosyncracies are generally prohibited from surfacing by the usual administration of the instrument, which limits each of the respondents to a single check mark on each of the scales.

Early in the administration of the Ideal Scientist SD, a number of the scientists asked if they could give multiple responses to some of the scales. The respondents said they found it extremely difficult to give a single check mark for many of the scales. Instead

116

of a single check mark, they felt that a range of responses was called for. The ranges expressed the fact there were a range of possible scientific types that were compatible with the concept of the Ideal Scientist. Thus, for example, a number of the respondents drew a line across the whole warm—cold dimension, indicating that all types were equally possible. To say the least, this wreaks havoc with some of the usual statistical techniques (like factor analysis) for analyzing the data. However, one of the things I was interested in studying was precisely how the respondents viewed and reacted to the scales. And so I did nothing to discourage this kind of scoring behavior. Indeed, I incorporated it as part of the standard instructions to the rest of the sample. I informed the respondents who were interviewed later that if they wished they could use multiple checks if they so felt a particular scale called for it.

As behavioral scientists, one of the things we should be interested in is how sophisticated respondents view a difficult concept. All too often we let the ease with which we can analyze certain kinds of responses determine the kinds of responses we will allow our subjects to manifest. The question is: Which is to control which? Should the actual responses control the instrument, or should the instrument control the form of the responses collected? The answer depends on many complex factors, such as what one wants out of the research, and ultimately on one's philosophy of research. There is a lot to be said for sticking with the tried and the true, the traditional, well-defined methods of analyses. The prime benefit of that approach is that one has clear rules to follow for analysis and interpretation. On the other hand always sticking to the tried and the true may inhibit the development of new methods. It is my philosophy of research that, to prevent inhibiting the development of new methods, one should continually venture out from the commonly accepted by asking such questions as: What would happen if we relax this assumption? this constraint? Such was the intent behind permitting the multiple responses to emerge.

I will say more about the marked reaction of the respondents to the warm-cold dimension after we review the responses to all of the scales. For now, it is merely worth noting that the judgment of irrelevancy tells us something very interesting about the psychology of scientists and provides an interesting linkage to the previous literature in the area. The mixture of different kinds of adjective-pairs in a semantic differential enables us to study how the respondents organize their concept space. I am explicitly interested in what the respondents regard as important to the concept being scaled versus what they regard as unimportant or inessential.

Scale 3: Consistent—Inconsistent. Some representative responses:

Scientist Nolan: It's important for a man to be very highly capable of logical reasoning, and in this sense, consistent. But it is not necessary for him to behave always in the same way. Many times people work in bursts and this kind of inconsistency is more than acceptable. The greatest scientists only make a few great discoveries in their lives, and their behavior at those times is quite different from their behavior at other times.

Scientist Oaks: The Ideal Scientist should not be totally consistent or he won't be able to follow his intuitions.

Scientist Park: Obviously it's desirable to be consistent, but a scientist somewhere in between might be more the ideal than one who is completely consistent. I don't object to anything that appears inconsistent to the outside world as long as it's all right with me.

Scientist Quay: You can say that Bracken [an important scientist] is tremendously consistent. In [1970] he still believes what he came up with in [1950]. This is consistency and rigidity. A man should be consistent in the sense that a theory he proposes in September should be consistent with what he proposed in March, and if not, he should say so. This is a matter of integrity. A man shouldn't look at one aspect of the picture and come up with a theory, then look at another aspect another time, and forget what he said before.

Scientist Reed: Ideally you should be completely consistent, but you shouldn't defend an idea after it becomes untenable. Young is capable of this.

The comments of Scientists Quay and Reed are particularly interesting because they indicate that the scales are not being responded to as isolated entities. Strong background beliefs and whole concepts about science and scientists come into play. Thus, Scientist Quay links the concept of consistency with that of rigidity. And both Quay and Reed, of their own accord, bring up the names of specific individuals and use them as points of reference. No matter what the exercise and no matter how abstract and seemingly far removed from individual personalities, time and time again the respondents brought up the names of specific individuals. To put it mildly, the images of particular individuals were very well fixed and very near the response surface.

118

Scale 4: Dull—Brilliant. Some representative responses:

Scientist Wilson: You can't have them [scientists] dull, but you can have them less than brilliant. You need some dull types to plow through the dull parts of a research problem.

Scientist Xilas: I obviously lean toward the brilliant end but a man can be so brilliant in one subject that he doesn't have a well-balanced outlook toward science in general or toward science and its relation to the rest of life. Being too brilliant can be dangerous in some cases. I'd check between 5 and 6.

Scientist Young: There is a need for dull scientists. But since you're asking for the Ideal Scientist I'll check the finest degree of brilliancy.

Scientist Zimmer: Obviously you want them as brilliant as possible. I can't see any justification for being dull if you can avoid it.

As I remarked in the preceding chapters, no matter what the issue, for every scientist who took one position there was another scientist who took the opposite. Scientist Zimmer cannot see any justification for having dull scientists; Scientists Wilson, Xilas and Young can see many.

Scale 5: Experimentalist—Theoretician. Some representative responses:

Scientist Baker: A good theoretician must have a good understanding of experiments and those who do experiments need a good theoretical background. I'd say this scale runs from 4 to 7 as you can't just be a plain experimentalist.

Scientist Case: In a funny sort of way, I consider myself a student of Fermi, who was closest to being a great experimentalist and theoretician. An Ideal Scientist should be right smack in the middle. An experimentalist with a little knowledge of theory can do some great clever things; a theoretician with a very modest knowledge of experimentation can do some very brilliant things. Einstein was a clod in the lab. So there is not an ideal.

Scientist Davis: You can have the entire range here.

Scientist Edwards: The whole range is possible.

Scientist Ford: The Ideal Scientist has to be both simultaneously. I don't know how a scientist becomes one or the other. Sometimes an experimentalist will later become a theoretician, or vice versa. A theo-

119

retician may require very specific data no one else will give him, and have to do the necessary experiments. Sometimes a theoretician has to collaborate with an experimentalist. The Ideal Scientist ought to be aware of all the theoretical ramifications of certain theories. Some scientists compile data and do not interpret it; they might not even qualify as scientists. X is a compiler, a walking encyclopedia. Compiling is a necessary part of the scientific effort. The other type of scientist thinks in terms of broad problems as well as the current problems in the field of science. They stay at the forefront, and make the most significant contributions. It's difficult to say who these people are. We need Adams but not Smith, who does not have a regard for scientific facts. Jones and Ford are examples of theoreticians who, like Hall, are not familiar with geological data. A theoretician studying the moon must be familiar with a wide spectrum of disciplines. I'd check both 1 and 7 simultaneously.

Scientist Gage: This is where you pick up people's hang-ups about what they think is important. I think both are very important, so I'll go right in the middle with a 4.

Scientist Hall: The Ideal Scientist would be exactly half way. He'd be good at experimentally collecting data, but would have an equal ability to interpret it theoretically.

This scale is the first of the *general cognitive style* dimensions (scales 5, 6, 9, 20, 21). The dominant characteristic of these scales is that, of all the dimensions, they elicited the largest number of bimodal responses — a check in both the 1 and the 7 spaces simultaneously. The bimodal response to these scales was so strong that a significant number of those scientists who gave a single check mark in response to all of the other scales responded with a check in both the 1 and 7 spaces on these particular scales. Of the 45 scientists who responded to the SD, twelve gave 4's to scale 5, seven checked both 1 and 7, and six checked the whole range from 1 to 7. The difference between those who checked 4 and those who checked 1 *and* 7 is one of interpretation of the notion of *equality of the traits*. Those who checked 4 generally felt that their Ideal Scientist should be equally in the middle, whereas those who checked 1 and 7 felt their Ideal Scientist should possess both end qualities of the scale *simultaneously* or be able to move back and forth between each of the extremes at will. There is a difference between the two; one (the 4 response) is a scientist who is a blend of the extremes whereas the other (1 and 7) has the capability of taking on, at different times, both extremes to an equal degree.

An inspection of Figure 2 shows that the mean for every one of

120

the scales in this category (5, 6, 9, 20, 21) is very close to 4, whether it was produced by an initial score of 4, 1 and 7, or 1 through 7.

Scale 6: Practical—Theoretical. Some representative responses:

Scientist Ingram: Both are needed.

Scientist Jones: This is wideopen. The only good theoreticians I know are also practical-minded so I'd have it going from theoretical down to 2.

Scientist King: The middle part of the scale is not quire so useful as the extremes are. The experimentalist who had no understanding of theory is not very useful. A theoretician who is not an experimentalist at all is nevertheless useful.

Scientist Logan: The whole range.

Scientist Nolan: This is asking the same question as before; both sides are required.

The importance of the responses to this and the preceding dimension will become even more apparent in the following chapter, where I discuss some of the attitudinal items that dealt directly with such issues as the relationship between scientific theory and data.

The responses to scales 5 and 6 were a preliminary indication of the fact that, time and time again, the respondents stressed the strong interdependence between theory and experiment. Not only was a good theory dependent upon a good experiment, and vice versa, but to be a good experimentalist one had in some sense to be a good theoretician, and vice versa. Contrary to a theory-before-data or a data-before-theory view of science [72] discussed in Chapters Five and Seven, scientific observations were not viewed as theory-free or theory-independent.

Scale 7: Aggressive—Retiring. Some representative responses:

Scientist Oaks: In one way or another the really top people are aggressive; they push for their ideas.

Scientist Park: The answer to this is meaningless. Scientific behavior is separate from human behavior.

Scientist Quay: A scientist should be moderately aggressive as he's often the only proponent of a new idea, and unless he's a little pushy and single-minded about it, a good idea may be abandoned too soon. But he shouldn't be so aggressive that he makes life difficult for others.

Scientist Reed: I can't see any advantage or use to being retiring. I'm retiring but not by choice. This isn't something you would wish on a person. You want people to listen to your ideas, to listen when you tell them to do something, or ask for support. All it takes is aggressiveness. I wouldn't mark this too far on the aggressive side however, because when you get that aggressive you begin to be like scientist X, offensive.

Scale 8: Amoral—Moral. Some representative responses:

Scientist Smith: I have to say moral. I don't know how anybody could say anything else.

Scientist Tate: I'm getting confused between the Ideal Scientist and Ideal Person. I have a strong bias toward people being moral but I've seen some outstanding scientists who are completely amoral.

Scientist Upsher: Come on, Mitroff. This thing is absurd. You're asking me if it's all right for a guy to be moral only 10 percent of the time.

Scientist Victor: This has to do with people. This has very little to do with science.

Scientist Wilson: I can't answer. In most of my experience, scientific problems deal with morality very little. I can't think of a specific case where any of this has anything to do with morality because this science [geology] deals mostly with the inanimate part of the living world.

Scientist Xilas: This doesn't matter as far as science is concerned. But in a way it does at least as far as the stealing of the work of others goes. Morality is an important thing. I ought to be strong on that. My work was not really even stolen, just on a very minor scale. I know of cases where people have been stolen from, and it's a great scar. I think stealing is mostly an unintentional thing. Two years ago Baker visited my lab. My colleagues and I gave him permission to write about one aspect of our work in a short paper in XX. Baker came out with one of the longest papers ever in XX spewing out most of the stuff we had given him. It's well known that King gave Baker his ideas on P. For such an enormous new concept it's very sad that a man like Baker is quoted for this. People send Baker stuff just for him to study, and you'll see them in papers and journals. This is very rare, though. I think Baker is just an ambitious son of a bitch.

Scientist Young: I would go most of the way towards moral. However, there are times when you have to step on people. If you are completely moral and have other people's welfare at the top of your mind, you would compromise your science occasionally.

Scientist Zimmer: There is no excuse for stealing results. Amoral could mean that you do science regardless of the results you get.

The responses to this scale are interesting, not only for what is explicitly said but for what is not said. The prevalent sense in which the respondents conceive of scientific morality, if they grant it a place at all, is within the rather specific and limited context of stealing. Although this is certainly an important context it is by no means the only, let alone the most important, sense in which morality bears on science. Indeed, it can be argued [72, 79, 81] that moral judgments penetrate to the very core of the scientific enterprise, the validation and conceptualization of scientific knowledge and statements. There were few indications that the respondents had grappled in any penetrating and deep way with the relationship between so-called scientific *statements* and ethical or moral *judgments* in this sense (see Chapter Seven). A few of the respondents were seriously concerned with the relationship between their science and morality in the sense of the larger society. But even here, with the exception of a very few individuals, the discussions indicated that the respondents had engaged in little systematic and sustained study of the complex issues involved. Although this may be understandable, given the press of traditional scientific demands on their time, it is less defensible now, if it ever was. In an age where complex issues and problems cut across and blur the distinctions between disciplines even more, we can less afford to have isolated experts in science, ethics, and policy-making. When scientists deal with problems that by their very nature involve all three areas, we need scientists whose conception of morality is broader than the limited context of stealing.

As a side comment, the responses of this section indicate that stealing, however rare its actual occurrence may be, is more than just an isolated topic of concern. The scientist of the last chapter who was concerned about stealing is not an isolated exception. If he differs from the respondents of this chapter, it is only with respect to his overwhelming preoccupation with stealing.

Scale 9: Rational—Intuitive. Some representative responses:

Scientist Adams: A great amount of intuition goes into a good scientist;

123

there is subconscious reasoning. Completely rational people don't make good scientists.

Scientist Baker: I do not regard these as opposites of a continuum. I think one can be both. Rationality applies to those cases where the information available is sufficient to make a conclusion; then good scientists ought to be strictly rational. But when the data are not sufficient to come to a definite conclusion, an intuitive person uses the vague or not strictly applicable information to best advantage.

Scientist Case: I don't believe there is such a thing as intuition. Intuition is just a large number of small rational steps which go by so fast you call it intuition instead of rational. The only justification for being an intuitive reasoner is that you hope to be rational about it.

Scientist Davis: I believe that objectivity and reason go hand in hand, but there are times when intuition is a remarkable tool. I wish I had more of my wife's intuition and less of my own tendency to be reasoning in my approach. I lean toward the intuitive side of this scale but with an obvious dependency on good reasoning processes.

Scientist Edwards: You need both very much [checks 1 *and* 7].

Scientist Hall: You should be rational when you're selecting and compiling data, and intuitive when you interpret it [checks both 1 and 7].

Scientist Ingram: You ought to be intuitive when you're designing an experiment and even when you're collecting data and compiling it, but rational when you're interpreting it. I'd mark both 1 and 7.

Whatever intuition meant to the respondents — and as the comments indicate it meant many different and contradictory things — one thing stands out: 23 of the respondents, or 51 percent of the sample, saw *rational* and *intuitive* as complementary forces of equal importance, not as opposites (10 scientists checked 4; 8 checked 1 and 7; 1 checked 1, 4, and 7; and 4 checked 1 through 7). Thus, the majority of the responses were bimodal or multiple. Of the remaining 20 respondents, 8 gave no response at all to the scale (blanks); 6 gave responses on the rational side; and 6 gave responses on the intuitive side.

The significant finding is the sample's appreciation that rationality and intuition must go hand in hand. Each sustains the other; each is necessary for the other. Rationality, logic, and systematic method without hunches and intuitions are empty, confining, and blind. Intuition without some form of rationality is wild, unintelligible, aimless, and undisciplined. In this sense, the sample's comments are a welcome counter to philosophical analyses of intuition. One of these is Mario Bunge's *Intuition and Science* [56], which presents intuition in the worst possible light, as inarticulate visions and hunches that never link up with rationality and pretend to stand above it. It is easy to analyze — rationally of course — such forms of intuition out of science. What is hard is to preserve intuition so that it both aids and is aided by science and rationality. Bunge writes: "Scientists esteem intuition, particularly creative imagination, catalytic inference... but do not *depend* upon it." He is right of course, but then neither do they depend solely on rationality either. Or rather, they depend on both.

Scale 10: Political—Apolitical. Some representative responses:

Scientist King: This is not a scientific question; it's a moral one. It doesn't have anything to do with science in terms of being a scientist. But I damn well think we need more political scientists who study politics.

Scientist Logan: Scientists should be politically aware of what's going on.

Scientist Meade: This scale is not really relevant.

Scientist Oaks: I don't have much regard for people who wash their hands of politics. They will be scientists first, but they are citizens also. Nobody can afford to be apolitical, especially when one grants that scientists are fairly influential. One shouldn't be at the extreme of the spectrum; I'd check something like 2 or 3.

Scientist Park: He should be slightly on the political side as certain large scientific projects like this one are not free of politics, like who gets what lunar sample. People on the selection committee always protect their own field of interest, and deprive many other people of critical samples. A lot of material was used up in analysis of P. Edwards and Ford were very influential on this committee. Nolan thinks nothing but Q analysis is important. Gage knows what "we" need in terms of samples. So I and others didn't get enough Apollo 12 samples to do a thorough analysis.

Scientist Reed: This goes with aggressiveness. One has to be political. One has to fight for what he wants.

The most interesting thing about the responses here is the surprisingly large number of scientists who perceived that science and politics were vitally linked together at a variety of levels. Politics affects and infuses science at the micro and macro level, and science affects and infuses politics. At the micro level, the make-up of the key internal scientific advisory committees directly affected the doing of science, and in this sense, affected the direct production of knowledge of the moon's physical properties (see the comments of Scientist Park). In this regard, *the determination of the moon's properties is not just a scientific problem but a political and social problem as well.* In this sense, I disagree strongly with scientist King's comments. The two aspects are not separable as he would separate them. Science is a social institution whose internal organization and internal politics affect its inquiry structure [72]. Time and time again throughout the entire study, the respondents bemoaned the fact that some of the scientists unfairly used their scientific-political influence to get this or that sample, push a particular experiment on board a particular Apollo mission, or force through the decision to land at a particular site on the moon. Whether this was literally true I have no absolute way of knowing, and as I mentioned in the last chapter, this is almost beside the point. The key thing is that so many scientists mentioned this and wanted to believe it. However, they also mentioned that "on the whole" they thought everyone was treated fairly and that no one point of view dominated the planning of the missions. Whatever the case, the point still remains that important decisions were hammered out in key committees which vitally affected the science of the missions. In this sense, science and politics at the micro level have become more and more inseparable, if science ever really was separate from politics even at the micro level.

At the macro level, the realization of the interaction was even greater. It was clearly acknowledged, and with more than just an infrequent tone of bitterness, that Apollo was basically a program that was initiated out of political motives, not scientific ones. Nobody was going to spend 25 billion dollars for science alone. Science was clearly a second-class rider in the program, but better a second-class rider than no rider at all. The general feeling was that without politics — or "geo-politics" in the respondents' jargon for any large-scale politics affecting their science — there would not have been any science at all. The very term *geo-politics*, which was used so freely, was an explicit acknowledgment that lobbying was an important part of the scientific enterprise.

Another indication of the appreciation of the strong tie between science and politics is provided by the responses to some of the additional questions that were put to the scientists. In response to the question of whether "scientists as scientists should take an active role in politics," 10 agreed strongly (SA), 18 agreed

(A), 2 were uncertain (?), 11 disagreed (D) and 1 disagreed strongly (SD). In response to the question of whether "more scientists should run for public office," there were 9 SA responses, 22 A, 9 ?, 2 D, and zero SD. Many of those who agreed (A and SA) with the questions felt that science needed expert, informed representation in Congress. Such representation was not only in the interest of science but also in the interest of the larger society. Unfortunately this is not the place to discuss the full implications of these and the other items or to compare them with some of the most recent literature in the politics of science [157, 249].

The first ten scales give a fair sampling of the various kinds of responses. They also give a fair sampling of the various kinds of scales: warm-cold (2), aggressive-retiring, (7), amoral-moral (8), and political-apolitical (9) represent *general personality variables*; impartial-biased (1), consistent-inconsistent (3), and dull-brilliant (4) represent *general scientific traits*; and experimentalist-theoretician (5), practical-theoretical (6), and rational-intuitive (9) represent *cognitive style variables*. Thus, although the responses to the remaining scales are just as interesting, they add little to what has already been reported. For this reason, they have been omitted.

General Conclusions

Given both the preliminary quantitative results (the profile of mean scale scores and the t-test results) and the qualitative or verbal responses, one can provide a more definitive response to the question of what the results imply for the acceptance or rejection of the HSS and HAS concepts. Table 9 presents a comparison between the implications that can be derived from the quantitative and qualitative results respectively. The upper half of the table shows the quantitative scale responses which support the HSS concept, those which support the HAS concept, those which support both concepts, and those which support neither concept. The bottom half of the table shows the same thing for the qualitative responses. In the upper half of the table, the criterion for "support of a concept" is one of "profile closeness"; for example, scale 1 tends to support the HSS concept because the mean sample response (2.75) on scale 1 is "closest to" the "impartial" end of the scale, which is an element of the HSS profile as defined in Table 7. In the bottom half of the table, the criterion is that of the "dominant tendency of the verbal responses" to the scale by the sample;

Table 9.

Rejection/Acceptance of HSS and HAS: A Comparison Between the Quantitative and Qualitative Scale Responses

Quantitative Scale Responses

Adjectives supporting HSS		Adjectives supporting HAS		Adjectives supporting Both		Adjectives supporting Neither	
(1) impartial	SP	(2) warm	GP	(5) experimentalist-theoretician	CS	(6) practical-theoretical	CS
(3) consistent	SP	(8) moral	GP			(9) rational-intuitive	CS
(4) brilliant	SP	(13) altruistic	GP			(10) political-apolitical	GP
(7) aggressive	GP					(20) speculative-analytical	CS
(11) self-confident	GP					(21) generalist-specialist	CS
(12) hard-driving	GP					(25) spontaneous-calculating	SP
(14) flexible	SP					(19) democratic	GP
(15) individualistic	GP					(23) liberal	GP
(16) ideals-oriented	GP						
(17) offbeat	GP						
(18) open	SP						
(22) skeptical	SP						
(24) creative	SP						
(26) diligent	SP						
(27) precise	SP						

128

Table 9 (continued).

Qualitative Scale Responses

HSS		HAS		Both		Neither		
(3) consistent	SP	(1) biased*		(5) experimentalist-theoretican	CS	(6) practical-theoretical	CS	
(4) brilliant	GP					(9) rational-intuitive	CS	
(7) aggressive	GP					(14) flexible-rigid	SP	
(8) moral ++	GP					(18) open-closed	SP	
(10) political $	GP					(20) speculative-analytical	CS	
(11) self-confident	GP					(21) generalist-specialist	CS	
(12) hard-driving	GP					(25) spontaneous-calculating	CS	
(13) self-serving	GP						SP	
(15) individualistic	GP					(2) warm-cold +	GP	
(16) power-oriented	GP					(23) liberal-conservative +	GP	
(17) offbeat	GP					(19) authoritarian	GP	
(22) skeptical	SP							
(24) creative	SP							
(26) diligent	SP							
(27) precise	SP							

Key: GP = General Personality Variable; SP = Scientific (Related) Personality Variable; CS = Cognitive Style Variable; * = "biased" in the sense that a scientist must be "committed"; ++ = "moral" in the sense of not stealing, otherwise "amoral"; $ = "political" in the sense of being "aggressive"; + = scales judged "irrelevant" to Ideal Scientist concept.

129

for example, scale 1 tends to support the HAS concept, because the overwhelming verbal response was that the Ideal Scientist had to be "biased" in the special sense that scientists had to have strong "commitments." For the most part, the quantitative and qualitative results are identical. The interesting comparison is therefore not between the quantitative and the qualitative results but between the columns of the table.

A detailed inspection of Table 9 is very revealing. Table 9 clearly shows that it is exclusively the general personality (GP) and scientific personality (SP) variables which support the HSS concept. None of the cognitive style (CS) variables support it. All of the CS variables support neither concept as they have been initially defined (see Table 8). If one were to redefine the HAS concept along the lines of the "Neither" column — that is, if the HAS concept were to be merely a strongly opposing concept (with scale values of 4 on the CS dimensions) and not a completely antithetical concept to the HSS in the sense of extreme opposing scores — then the HSS concept is supported or affirmed on nearly all of the personality dimensions, and the HAS would be affirmed on all of the cognitive style dimensions. In other words, *the sample supports the traditional stereotypical concept of scientists on the personality dimensions and rejects it on the cognitive style dimensions.* Or, because the ends of the personality scales which fall on the HSS concept are masculine personality attributes or virtues in a traditional and stereotypical sense of masculinity, we can say that *the sample supports the HSS concept in the sense of traditional masculine values (aggressive, hard-driving, self-serving, power-oriented, authoritarian, skeptical, diligent, precise) and rejects it in the cognitive sense (one needs to be both practical and theoretical, rational and intuitive, and so on).* The support of masculine values becomes even greater if we shift the warm—cold dimension from the Neither column to the HSS column. The overwhelming judgment of the irrelevancy of this dimension can be interpreted as a suppression of affective responses or feelings, a finding that is consistent with previous research on the psychology of scientists [285]. That is, the suppression of or the relative unconcern with feelings is a masculine characteristic in a traditional sense of masculinity.

This split in the dimensions[4] helps explain why the responses are so revealing on the CS dimensions and so commonplace on the personality dimensions. The CS responses are novel, unexpected, enlightening; the responses on the personality dimensions are typical. They sound like a recitation of the typical American virtues, the typical male pattern for success: hard work, dedication, striving, and even a touch of ruthlessness (self-serving, authoritarian). When one thinks about it, none of this is really surprising. We are dealing with a sample that has attained more than its share of success in the traditional sense of power (access to people and money), status, and prestige. The responses merely affirm what it probably takes to achieve success in any male-oriented and dominated profession. In this regard, the adjectives under the HSS column in Table 9 do not apply exclusively to science any more than they do to any male-dominated profession.[5]

If all this is an accurate representation of the beliefs of scientists, then it is not surprising that science has attracted so few women. To make it big in such a system, instead of just being allowed to serve in subservient or support roles, would require women who did not match the traditional and conventional concepts of woman [117]. I am not saying that this is as it should be or that this is the only way science can or should be sustained. However, I am saying that a sample of persons who have achieved success have the kinds of values that are compatible with and reflect a very particular form of success. Further, if the system tends to attract and self-select those who believe similarly, then the system becomes close to self-sustaining, a hard nut to crack and to change. Apparently it is easier to tamper with and even to change scientists' cognitive image of science (so scientists can be both rational and intuitive) than it is to change traditional masculine values. Science truly is a masculine way of looking at the world.

Preliminary Analyses of Individual SDs

The respondents' images of the Ideal Scientist were filled with some equivocation. But there was virtually no equivocation in their images of the particular individual scientists. To say that their images of their fellow scientists were firm and decisive is putting it too mildly. At least this was true for the particular subset of 10 scales that was selected for use in the rating of the particular set of individual scientists. Only 10 scales were used for these individuals instead of the full 27 in order to reduce the time used for administering the Individual Scientist SDs; 27 scales would have consumed too much time and made the exercise too

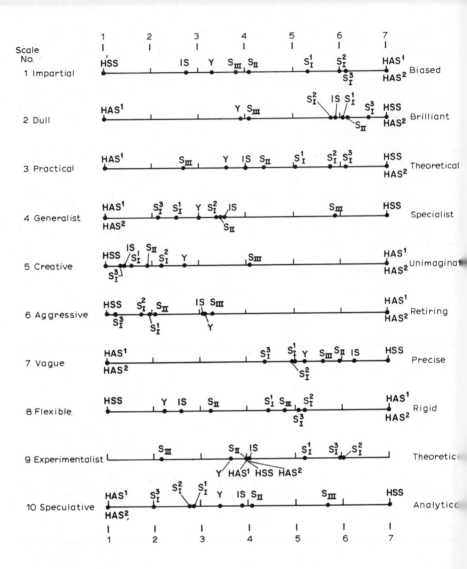

Figure 3. SD profiles of individual scientists. S_I^1, S_I^2, S_I^3 are three representatives of highly speculative-theoretical scientists; S_{II} is representative of the excellent combiner of theory and experiment; S_{III} represents the average experimentalist; Y is the general category of all respondents (for example, "yourself"); IS is the ideal scientist.

132

tedious, even though it would have been extremely interesting.

Figure 3 presents the profiles for each of the Individual Scientist SDs. That is, Figure 3 presents the means of the sample's perceptions of each of the specific individuals who were picked as representative of the different types of scientist discussed in the last chapter. S_I^1, S_I^2, S_I^3 represent the three individuals who were nominated most frequently as representative of the highly speculative, theoretical type of scientist. S_{II} represents the individual scientist who was nominated most frequently as representative of that type of scientist who was an excellent combiner of both theory and experiment. S_{III} represents the individual scientist who was nominated most frequently as representative of the average scientist and as a good to outstanding but narrow experimentalist. Y represents the general response category Yourself; every scientist also filled out an SD profile on himself in addition to the SDs on S_I^1, S_I^2, S_I^3, S_{II}, and S_{III}. HSS represents the Hypothetical Stereotypical Scientist and HAS^1 and HAS^2 represent two versions of the Hypothetical Anti-Stereotypical Scientist. Finally, IS represents the Ideal Scientist mean scores plotted for the particular set of ten scales.

Figure 3 contains a great deal of information that is highly significant, statistically as well as psychologically. Concentrating on the Individual Scientist points, the first thing one notes is that there is a great deal of regularity to the placement of the points on the individual scales. The various types do not fall randomly on the scales. There is a definite pattern to their order of occurrence. If we consider the location of the points on each scale as judgment rankings by the sample, and if we realign some of the scales to compensate for the random order in which the ends of the scales have been arranged so as not to bias the subject responses, then we can compute a Kendall coefficient of concordance W [288, pp. 229—238], which measures the overall agreement among the 10 sets of rankings. For the particular set of points, W = 0.738, which says that the agreement between the orderings on the separate scales is quite good (a W of 1.0 would indicate perfect agreement). Another way to put this is to say that the pattern $S_I^3 - S_I^2 - S_I^1 - S_{II} - Y - S_{III}$ is a good general representation of the ordering on each of the scales. Indeed, it can be shown that the above pattern emerges from a variety of separate analyses and is thus significant. The fundamental importance of the pattern is that it indicates the basic ordering in the sample's perceptions of the various types. It also gives a separate and needed validation of the definition of the various types. For each scale, the three type I's are bunched closely together, followed by the type II scientist. This is comforting, for this should be the case from considerations of face validity. Next comes the general category Yourself, followed by the type III scientist, which on a number of the scales is quite far from Y. This, too, is comforting for, from the verbal comments of the respondents in round I, Y should be between S_{II} and S_{III} and closer to S_{II} than to S_{III}. The terms describing S_{II} were more favorable than those describing S_{III}. On every scale, the distance between S_I^3 and S_{III} is considerable. Indeed, S_I^3 and S_{III} mark the extremes of the scales in nearly every case. This, too, should be the case from general considerations of face validity involved in the basic definition of the types. To say the least, we have a graphic portrayal of what the types look like on various scales and where they stand in relation to one another.

It can also be shown that the distances between the Individual Scientists on each scale are significant. A one-way analysis of variance of the sample responses for each scale indicates that the differences are statistically significant. For each scale, the general difference between all the types is significant at an α level far below the 0.001 level, that is, $\alpha \ll 0.001$. In other words, the differences between the types perceived by the sample were highly significant on each scale. We are not dealing with weak or trivial differences between the types or their representatives. *The differences in perception between the types are extremely strong.*

Table 10 presents an adjective profile comparison of the various Individual Scientist types. There are a few things in Table 10 that deserve explanation. It may appear contradictory that the three type I scientists were perceived as highly creative and at the same time as rigid. Here again one has to look at the comments of the respondents to understand the sense in which the terms are being used. The three type Is were perceived as extremely creative in the sense of their being able to produce and having produced many original, innovative ideas over a long period of time. In this sense, they were regarded as extremely flexible. They possessed the requisite mental agility and nimbleness of mind to see old problems in a new light and to perceive (literally invent) highly imaginative patterns in complex sets of data; thus, the judgments that they are *brilliant, theoretical, generalist, speculative,* and the tendency towards *vagueness.* On the other hand, they were perceived as extremely attached to their ideas once those ideas were produced; thus, the strong judgments of *bias, aggressiveness,* and *rigidity.*

Table 10.

Adjective Profiles for Scientific Types

Scale Types

	S_I^3	S_I^2	S_I^1	S_{II}	Y	S_{III}
1	Markedly Biased	Markedly Biased	Moderately Biased	Neither or both Impartial-biased	Moderately Impartial	Neither or both Impartial-biased
2	Extremely Brilliant	Markedly Brilliant	Markedly Brilliant	Markedly Brilliant	Neither Brilliant nor Dull	Neither Brilliant nor Dull
3	Markedly Theoretical	Markedly Theoretical	Moderately Theoretical	Both Practical and Theoretical	Both Practical and Theoretical	Moderately Practical
4	Markedly Generalist	Moderately Generalist	Markedly Generalist	Moderately Generalist	Moderately Generalist	Markedly Specialist
5	Extremely Creative	Markedly Creative	Markedly Creative	Markedly Creative	Moderately Creative	Neither Creative nor Unimaginative
6	Extremely Aggressive	Markedly Aggressive	Markedly Aggressive	Markedly Aggressive	Moderately Aggressive	Moderately Aggressive
7	Neither or both Vague-precise	Moderately Precise	Moderately Precise	Markedly Precise	Moderately Precise	Markedly Precise
8	Moderately Rigid	Moderately Rigid	Neither or both Flexible-rigid	Moderately Flexible	Markedly Flexible	Moderately Rigid
9	Markedly Theoretican	Markedly Theoretican	Moderately Theoretican	Both Experimentalist-theoretican	Both Experimentalist theoretican	Markedly Experimentalist
10	Markedly Speculative	Moderately Speculative	Moderately Speculative	Both Speculative-analytical	Moderately Speculative	Markedly Analytical

Note: Extremely or significantly = 1.0 to 1.5 or 6.5 to 7.0 scale values on SDs; markedly = 1.5 to 2.5 or 5.5 to 6.5 scale values on SDs; moderately = 2.5 to 3.5 or 4.5 to 5.5 scale values on SDs; neither or both = 3.5 to 4.5 scale values on SDs.

135

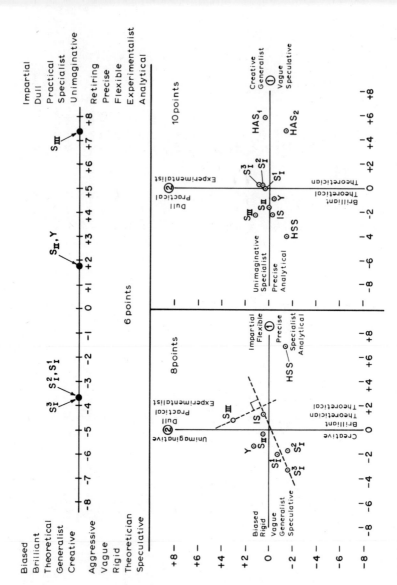

Figure 4. Three multidimensional scaling analyses.

Finally, Figure 4 presents the results of three multidimensional scaling analyses of the data contained in Figure 3.[6] (Three analyses were performed because multidimensional scaling is sensitive to the number of points being compared or configured [227].) Figure 4 is a vivid and graphic portrayal of where some real scientists and some hypothetical scientists lie in relationship to one another. The figure shows at a glance which kinds of scientists lie closest to which concepts. The figure also shows how far removed some real scientists or scientific types are from some hypothetical images of scientists, whether those images are of the Storybook or Anti-Storybook variety. It also shows again the stability of the pattern $S_I^3 - S_I^2 - S_I^1 - S_{II} - Y - S_{III}$. And in fact a separate multidimensional scaling exercise which was presented to the respondents before they took the Individual Scientist SDs shows that the same pattern $(S_I^3 - S_I^2 - S_I^1 - S_{II} - Y - S_{III})$ results from a completely independent exercise and not merely from an alternate way of analysing the same Individual Scientist SD data in Figure 3. We are dealing with well-defined and persistent differences between the types which appear invariant no matter what the analysis.

Summary and Concluding Remarks

With regard to the substantive results of the chapter, it is important to emphasize the nonrandom and systematic nature of the profiles that define both the Ideal Scientist and the Individual Scientists or types. Tables 9 and 10 are the clearest and the most instructive in this regard. Although there may not be any inherent, logical reasons why an experimenter (particularly of the S_{III} variety) cannot be any combination of the scale attributes, there are apparently strong psychological reasons for the respondents which limit the possible combinations of attributes. In other words, from a logical point of view there may be no strong a priori reasons for the interdependence of some of the scales, but from a psychological point of view there are apparently very good reasons, at least from the point of view of the respondents. Table 10 and Figure 3 indicate extremely strong, extremely well-defined differences between the kind of scientist who is perceived as a bold speculator (S_I), the kind of scientist who is able to combine both theory and experiment in creative ways (S_{II}), and the kind of scientist who is perceived as an experimentalist and representative of the average scientist (S_{III}). Table 10 and Figure 3 constitute a much more systematic and precise definition of the types than the descriptions in the preceding chapter.

The results of this chapter are obviously based on the basic judgments of a special group of experts and not on a formal analysis of the attributes of scientists. In other words, a group of peers is the basic measuring instrument; their individual perceptions constitute the basic data. The methods of analysis described constitute a set of techniques for transforming individual judgments into a kind of group judgment. The method of transformation most commonly used or at the heart of every technique is that of *averaging*.

As a consequence, the results of this chapter exhibit all of the strengths and weaknesses of systems that are dependent on the transformation of individual judgments. On the positive side, the respondents themselves have been used to measure specific combinations of attributes because of their presumed sensitivity to specific issues, a sensitivity that presumably extends beyond the capability of current purely formal methods of analyses to measure or derive (for example, logic). Even more important, one uses a group of scientists as I have done because one of the most important sources of information regarding science is what a group of practicing scientists think about science and their fellow scientists. On the negative side, the results may be confined largely to the particular group of scientists under investigation. The results may not generalize to other scientific groups.

It should be stressed that the perceptions of the representatives of the various types of scientists are truly those of the respondents. They are not necessarily mine. To take one example, in my many conversations with the individual designated as S_{III}, I very carefully looked for signs of the attributes attributed to him by his colleagues. In terms of his verbal responses alone, S_{III} does not justify the characteristics attributed to him. And in fact quite the contrary is true; in my individual judgment, S_{III}'s verbal responses were far more reflective and the antithesis of many of his colleagues' responses. But then this only confirms the crucial role that the judges play in methods of analysis that are dependent on judgmental scoring schemes. My judgments were based primarily on what I could observe and what I was sensitive to. In terms of this standard, S_{III} was superior in my eyes to many of his colleagues. In terms of the standards of his peers, he was not. I primarily judged his verbal and his attitudinal responses, and from these I tried to make an inference to his scientific work and work attitudes. His peers apparently judged both his work and his attitudes in other terms. I am not implying that the judgments of different judges ought to agree but rather I am noting that my assessment of S_{III} is not the same as that of his colleagues.

Leaving aside for later some of the problems connected with modes of inquiry that are dependent on the uses of judges, and especially the issue of whether one can legitimately extend the results of this study to other scientific populations, one of the most important issues to which we can address ourselves is the relationship between the results of this study and those from previous research in the area of the psychology of science. Three of the findings in this chapter are of particular importance: (1) the responses to the Ideal Scientist questions reveal a general split between the affective dimensions, such as warm—cold, and the cognitive dimensions, such as rational—intuitive; (2) the responses to the cognitive dimensions reveal that both ends, or extremes, of the scales are valued equally; and (3) the sample tended to accept the HSS concept on the personality dimensions and rejected it on the cognitive dimensions, and furthermore, the particular form of their acceptance of the HSS concept revealed a strong masculine orientation toward the world. Appreciation of the significance of these results necessitates a review of some of the previous research.

In a psychological study that can only be termed brilliant (*Contrary Imaginations* [203]), Liam Hudson identified and extensively analyzed the differences between two distinct kinds of conceptualizers: *convergers* and *divergers.* On a variety of diagnostic tasks, ranging from close-ended tests for analytic abilities such as most IQ tests measure to open-ended creative exercises such as the number of unusual uses that one could think of for ordinary everyday objects within some time limit, the differences between these two types stand out in extreme contrast. Convergers prefer to work on manageable, well-defined problems for which there exists a single best or right answer. Out of a set of many possible alternatives, they tend to converge on one and develop it in elaborate detail. Divergers prefer to work on (and invent) ill-defined problems for which there exist many possible alternative approaches, not solutions. Divergers prefer to multiply alternatives and possibilities rather than to develop any single one in great detail. From a single stimulus (problem situation, test item) they produce significantly more open-ended and alternative responses than convergers. Convergers are analytical; they are parts-oriented;

they are extremely skillful and adept at breaking complicated wholes into parts. They perceive systems and problems in separable [72] terms. Divergers are synthetic; they are whole systems-oriented. They tend to perceive systems as nonseparable [203, 310]. Divergers also tend to perceive problems, systems, and issues in value-laden terms. They tend to the pejorative and the personal in their argumentative and conceptual style. Convergers, on the other hand, tend to perceive and to conceptualize problems, systems, and issues in value-free and technical terms. They tend to the impersonal in their argumentative and conceptual style. As one might expect, these differences carry over to the ways in which these two types organize their total lives — that is, the ways they relate their professional to their personal lives [see 203, p. 93].

Hudson's studies [203—205] of English school boys show that a significantly large percentage of those classified as convergers tended to go into science and to become scientists, whereas those classified as divergers tended predominantly to go into the arts and the humanities. Walberg [456] traces the childhood origins of these two; Hitt [190] and Maslow [280] shed further light on their defining characteristics and draw their implications for the nature of science. In an important paper, "The Essential Tension: Tradition and Innovation in Scientific Research" [294], Thomas Kuhn argues that it is not an either/or, that both convergent and divergent attitudes are important for the advance of science and further that it is not necessary that both attitudes be concentrated within a single individual as long as both are present within the system in different individuals. It should really go without saying that this single dimension is hardly sufficient to exhaust all the shades and nuances of the human personality, nor is it meant to be. The convergent/divergent distinction is merely meant to indicate one important continuum of the human personality. As Hudson notes, most people are in the great middle of the scale; most people are blends of the two; relatively few individuals are found at the extemes.

By far the best and the most extensive summary on the psychology of scientists is in David McClelland's excellent paper, "On the Dynamics of Creative Physical Scientists" [285]. Drawing on the results of a number of other studies, McClelland shows that successful physical scientists tend to possess particular characteristics. His summary [285] is so good that it is worth quoting extensively:

> *Men are more likely to be creative scientists than women.* There are

no women among Anne Roe's [366—369] eminent scientists, and very few in *American Men of Science.* No fact is more obvious than the differential yield for science of the two sexes...

Experimental physical scientists come from a background of radical Protestantism more often than would be expected by chance, but are not themselves religious... [Roe] found that her scientists were not personally religious, a fact supported in Terman's [436] report that his scientists were less interested in religion than any of his other groups of comparable intellectual ability. In other words, scientists appear to come more often from a radical (thoroughgoing or strict) Protestant background and to reject it for science as a "way of life."...

Scientists avoid interpersonal contact. They are less gregarious, more autonomous, prefer working with things to working with people. Evidence for this generalization comes from many sources [including] McClelland [286], Cattell and Drevdahl [64], Cattell [65], Stein [420]...

The unsociability of scientists appears as early as age ten according to Terman's [436] data... Outstanding scientists are anything but normally gregarious. They like being self-sufficient and like being alone, probably because people and human relations seem both difficult and uninteresting to them...

Creative scientists are unusually hardworking to the extent of appearing almost obsessed with their work. Roe [366—369], in reporting on her eminent scientists, remarks that the *one* characteristic all of them seemed to have had without exception is an intense devotion to their work [*see also* 243]. There was never a question of putting in so many hours a day, a week or a year. Instead they worked nights, week-ends, holidays, all the time. In fact she wondered how they ever found time to be with their wives and families. Terman [436] also reports that his physical scientists stated more often than any others that their work itself gave them the greatest satisfaction in life...

Scientists avoid and are disturbed by complex human emotions, perhaps particularly interpersonal aggression. By its very nature science as an occupation glorifies objectivity, dispassionateness, or the impersonal search for truth. While personal biases and feelings sometimes crept into scientific work, ideally they have no place in it. For most scientists avoidance of human emotion is far more than simply an ideal as far as their profession is concerned; it runs as a theme through much of their thinking in other areas of life. Knapp [232] in particular found that science majors tell TAT stories significantly low in dramatic salience, in aggression, guilt or vindication, and in the tendency to bring the plot to a clear and decisive conclusion...

Actually one of the most striking things about the way the eminent scientists reacted to the TAT is their marked dislike for the test. The test requires a response to a number of dramatic human situations, and the scientists reacted by trying more or less strenuously to avoid responding to them at all in the usual manner. They found it extremely difficult to empathize with the characters pictured and to tell a dramatic story as they had been instructed to do. Instead they tended to block, to analyze various portions of the picture, to consider various

141

possibilities of action, and to be unable to decide on any one of them. The following initial reaction of one physicist to the first card (boy with violin) is typical: "That is most objectionable. We will carry out an analysis. I have all sorts of blocks because people are so unreasonable it always makes great difficulty for me.".....

To oversimplify a little by way of summary: *scientists react emotionally to human emotions and try to avoid them.....*

Physical scientists like music and dislike art and poetry...

Physical scientists are intensely masculine. On all interest and attitude scales that differentiate between men and women, physical scientists score very high on masculinity...

Physical scientists develop a strong interest in analysis, in the structure of things, early in life...

That the nature of this [early] interest is analytic is almost self-evident from the goal of physical science; this goal in the words of one of Roe's theoretical physicists, is to discover the "connections of things," and to get at the, "inner secrets of the world.".....

What is required of them in their profession has been enthusiastically adopted by them in their entire attitude toward life. For example, Terman's [436] physical science research group is highest in its interest in photography as an avocation, and photography is the method *par excellence* by which one "freezes" the flux of reality so that one can get a good look at its structure. As has already been suggested above, the task of telling stories to TAT cards was found to be particularly baffling by Roe's eminent scientists. The difficulty arises not only from a desire to avoid talking about human emotions, but also from their obsessive concern with analysis or with what is "really" a *correct* interpretation of the pictures...

They simply cannot let themselves go, in telling a tale of dramatic action, but are constantly brought up short by details which seem to require analysis and which do not fit into the [projective] story that has been started.

These, in brief, are some of the more important findings from previous studies. In general, the results of this and the previous chapter support these earlier findings.[7] However, the support is not complete or one-sided. It depends on which exercise we are looking at the results from or emphasizing at the moment. An even better way to put this is to say that the results not only reveal a tension between two opposing forces but a blend as well.

It is far too simple to say that the scientists of this study are either convergers or divergers. No one is that simple or deserves to be treated in such a simplistic and cavalier manner. Everyone has sides that are convergent and others that are divergent. It is thus not only more equitable but also more accurate to point out the ways in which the scientists reveal the convergent sides of their personalities and the ways in which they reveal their divergent sides or aspects. This, of course, is Hudson's point too.

142

The fact that so many of the scientists of this study judged such fundamental affective or interpersonal dimensions as warm—cold as irrelevant to their concept of the Ideal Scientist ("this has nothing to do with science") reveals a strong separation between their professional and their interpersonal lives. It is even putting it mildly to say that this is one of their strongest convergent aspects. On the other hand, the scientists revealed a marked divergent side of their personalities when they strongly insisted that it was not an either/or between analysis and speculation, rationality and intuition, and so on. The preference for both ends of the cognitive dimensions can be interpreted as a strong indication of a divergent aspect of their personalities.[8] However, again it depends on which results from which exercise we are looking at.

One of the most pertinent and interesting sets of responses in this regard occurred during the last round of interviews. One of the questions in particular produced a significant number of extreme convergent type responses. The specific question was, "If you could have any *one* (but just *one*) question answered tomorrow about the moon, what would that question be?" In posing this question I had nothing devious in mind. I was truly interested in what the scientists would designate as the unanswered or remaining questions about the moon that were uppermost in their minds. Thus, I was a bit taken back when up to 25 percent of the sample responded with something like, "Oh, you mean a feasible or an answerable question; there are lots of questions but they're either not answerable or likely to be answered, like the question of the 'origin' of life; therefore, I'll list a question which I think is likely to be answered or is answerable." What is so striking about this kind of response is that it reveals the imposition of constraints when none were implied by either the nature of the question (task) as stated or by anything in my behavior when presenting the question. In its original form, the question was completely open-ended; it was intended to be. It does not ask what feasible or answerable question one might ask. It also does not ask what question or particular type of question *I* meant them to ask. The question means what it says, or rather, it means what the respondents wanted it to mean, which of course is the point. It was their question to deal with as they saw fit. In other words, a significant percentage of the respondents took an essentially open-ended exercise and of their own accord turned it into a constrained exercise. Time and time again throughout the course of the study, this same pattern of behavior occurred in response to a variety of questions. In this sense, it must be concluded that there is a very strong converger or convergent side to the personalities of scientists. They are preoccupied with analysis, as McClelland notes. They cannot let themselves go, or at least a significant body often cannot, even when the freedom is given to them. In this regard, the results of this study support McClelland's generalization about the preoccupation of scientists with analysis. And the evidence for this generalization is even stronger here. McClelland's supporting evidence is based on projective TAT data, data that some might want to dismiss because it is far removed from the actual operating style and interests of scientists. The same cannot be done so easily here. We get the same kind of response McClelland observes in relation to a science-related task or exercise. That scientists *should be* analytically oriented is not the point. Of course they should. The question is whether they should be so analytically oriented or reality-bound that they cannot let go when it is required. As Kubie noted in the *Neurotic Distortion of the Creative Process*:

> The creative life in general makes extraordinary demands on the human personality. Specifically in science, creativity demands a flexibility which is perhaps greater than that of almost any other occupation to which man can dedicate himself. Precisely because the scientist must be as imaginative and free in fantasy as a poet, artist, or musician, and at the same time as tightly organized as the builder of bridges or the man who organizes and plans the split-second timing of an atomic explosion, we ask that the creative scientist should have a high degree of emotional and psychological freedom and imaginativeness coupled to an equal degree of organized precision. Few other occupations demand so much. I am not unhappy that science makes this demand of us. I am unhappy only at the fact that we do not recognize the implication of this demand and do nothing to help even the most gifted scientists attain the degree of emotional maturity and freedom which would make it possible for them to use their intellectual endowments in the most creative and constructive way that is possible for man to attain. I am unhappy at our complacency with a primitive educational process which to so large an extent reenforces everything neurotic in human nature [242, pp. 81—82].

The single result that stands out is the intense masculinity of these scientists. And indeed masculinity may be too kind or dignified a word. It is closer to the truth to say that it is their intense, raw, and even brutal aggressiveness that stands out. It is an aggressiveness that not only deeply infuses their relationships with one another but, as we have seen, their abstract concept of science. As one of the respondents put it, "if you want to get anybody to believe your hypothesis, you've got to beat them down with numbers; you've got to hit them again and again over the head with hard data until they're stupified into believing it."

If anything, the scientists of this study, unlike those of previous studies, displayed an over-aggressiveness. It cannot be said that they avoided aggression in any way. In this sense, they did not avoid complex human emotions. Again, however, we must note the discrepancies among the different types of emotions displayed freely. They were free and quick in displaying aggressive or harsh emotions; they were far less free, however, in displaying more affective or soft emotions. In this sense, they did avoid complex human emotions. They displayed only one half of the sphere of emotionality and that hemisphere with such an intensity that it tended to obliterate the other.

Chapter Five

Behavioral View of the Moon: Apollo 15

Chapters Three and Four were concerned with the exploration of the psychological and attitudinal differences among various kinds of scientists. More precisely, I have examined the perceptions of a particular group of scientists toward some of their colleagues and toward themselves. I also examined some of their attitudes toward some issues connected with science in the abstract. In carrying out this examination, there has been a constant movement between psychological and sociological issues. For example, in the preceding chapter it was shown how one could project some sociological conceptions regarding science, namely the conventional or traditional norms of science as represented in the HSS concept, onto a variety of psychological spaces. In other words it was shown how one could translate sociological concepts and variables into psychological ones and vice versa. The very act of performing this kind of analysis illustrates one of the major themes of this book — that the understanding of the nature of science and of scientists is a combined task for the philosophy, sociology, and psychology of science.

Examination of these issues is continued in this chapter, as is an issue that was briefly touched on in Chapter Three — the attitudes of the scientists toward some hypotheses pertaining to the problem of the moon's origin and their attitudes toward some other hypotheses having to do with the nature of the moon's physical properties. The underlying rationale behind this part of the analysis is that the understanding of the moon as a physical object is not only a problem for the physical sciences but also a problem for the social sciences. Because we ordinarily do not think of physical science problems as social science problems, it may be helpful to spell out more of what I mean.

The problem of understanding the moon as a physical object can be conceived of as a social science problem in the sense that social science techniques can be applied to the study of the attitudes of various individual scientists and groups of scientists toward various unsettled scientific issues, hypotheses, or theories. Scientific hypotheses, particularly unsettled or unresolved ones, can be viewed as attitudinal statements about nature toward which men can take a variety of stances. From this perspective, one can pursue a number of questions, such as: Do different types or groups of scientists have different positions on the issues? Do different groups change or shift their positions over time faster or more slowly than others? Do the answers to the immediately preceding questions shed light on the complicated phenomenon of how the personal judgments of individual scientists get transformed into collective, social judgments which then have the mark of disinterested, objective, universal physical truth? Put in this form, the understanding of the moon is most definitely a social science problem. The social sciences have a definite contribution to make to the understanding of how scientific judgments arise and influence one another.

The extent to which science is a problem for social science was brought home rather dramatically during the course of the interviews. A few of the respondents remarked that one scientist in particular was in the habit of checking out potential collaborators and that he had worked out a rather elaborate mechanism for accomplishing this. Whenever he spotted someone whose work might be of potential use to him in his own work, he would send out copies of that scientist's past work to trusted colleagues for their explicit and frank judgments. Apparently only if there was a strong consensus regarding the candidate's worth would that person's scientific data or arguments be relied on. Although this scientist's rating or evaluation scheme may be much more explicit than that of others, all scientists constantly assess the judgments, capabilities, and beliefs of their peers by the use of other peer judgments. Not all of these judgments can be construed or defended as purely scientific or rational in the traditional sense. If these nonrational or nonscientific judgments vitally affect the doing of science and the eventual content of scientific statements, then for this reason alone it is important that we study how and why these judgments get formed. The particular sciences whose explicit task it is to study these kinds of judgments are the social sciences. The study of science is itself a problem, a problem for the social sciences.

Public Opinion Poll

Table 11 presents the set of issues that were polled over time. Three judgments on each issue were required of each member of the sample. Two judgments were taken on interview round III; one of these judgments required the respondents to recall how they felt on the issues before the data of Apollo 11, and the other judgment required the respondents to indicate how they felt at the time of the interview, after the occurrence of Apollo 11, 12, and 14. One final judgment was later secured from the respondents on the last round of the study.

The judgments or attitudes toward each issue were expressed on a seven point scale. A judgment of 1 indicated strong agreement with an issue or hypothesis; a 7 indicated strong disagreement; a 4 indicated that one neither agreed nor disagreed with the issue. The first five issues were discussed in Chapter Three; they are some of the major hypotheses that have been proposed for the origin of the moon. The next two issues pertain to one of the fundamental controversies surrounding the origin of the moon, its thermal history and evolution, and even its present physical properties. This is the controversy over whether the moon currently is or has been hot to some substantial depth at some point in its history.[1] Short has briefly summarized the nature of the dispute and the evidence, much of it conflicting, that each side has appealed to in support of its position:

Running as common themes through all these hypotheses for the moon's origin are the ideas of accretion as the dominant formative process and the early proximity of the moon to earth. Most also assume some melting of the moon's outer layers and perhaps much of its interior. Thus, initially, the moon was *hot*, so that advocates of this view can claim a victory of sorts. The proponents of a *cold* moon can draw some comfort from its present status as an inert body with little or no internal magmatic activity. A rigid crust is evidenced by the mascons [discussed below] and the presence of an earthward bulge. A solid or nonmobile mantle is indicated by the low number of weak, internally originated moonquakes recorded by the seismometers. The absence of a magnetic field points to a nonfluid center (including at most only a small solid iron core). These all attest to the conclusions that temperatures in the lunar interior today are generally below the observable exterior.

One sure conclusion from Apollo sample analyses is that parts of the moon experienced some melting about 1 billion years after the initial crust solidified. Differences of opinion exist as to the cause and significance of this later period of magmatic activity. Questions still being

148

asked include: (1) Was this melting a general effect throughout the moon resulting from a gradual rise in temperatures or do the lavas come mainly from rapid heating in response to local or regional conditions? (2) Do the lavas originate from nearsurface or deepseated melting? (3) Could lavas represent residues of unsolidified liquid formed during a first melting of the moon shortly after its formation? (4) Is the melting initiated by heat buildup from radioactive decay, from impacts, from tidal friction, or from other sources of thermal energy? (5) Why are most mare lavas distributed on the frontside of the moon? [404, p. 231].

Closely related to the question of a hot or cold moon is the question of the cause or origin of mascons. First reported by Muller and Sjogren in *Science* [320], mascons are gravity anomalies. On the earth such gravity anomalies are usually related to positive or excess mass concentration over particular regions — hence, the term mascons. These are usually found over the highland or mountainous areas of the earth, where there are large buildups of mass. On the moon, just exactly the reverse occurs. The mascons are found over the low-lying regions like Mare Imbrium and Mare Serenitatis. The hypotheses listed in Table 11 are but two of the major explanations that have been advanced to cope with this phenomenon.

Finally, we come to one of the minor though at the same time one of the more fascinating of all the debates in the history of science, the origin of tektites. Tektites are circular pieces of glassy material scattered over some regions of the earth's surface — for example, Australia and parts of Southeast Asia. These objects are unusual and require explanation because the chemistry of the glass does not necessarily match the local chemistry of the region in which they are found. Thus, the question of their origin arises. There are two major theories of origin: (1) terrestrial, and (2) extraterrestrial. Those who favor a terrestrial origin argue that there are meteor or cometary impact craters, located hundreds of miles away, whose chemistry does match that of the tektites. Those who advocate this position have to explain how such small objects were transported hundreds of miles against the earth's gravity and air resistance. Ingenious mechanisms and arguments have been proposed to get around these difficulties. Those who favor an extraterrestrial origin offer equally strong and ingenious arguments as to why the moon is a more than likely source. Essentially this position gets around the dynamical difficulties of transporting small objects hundreds of miles through the earth's atmosphere by having tektites originate from meteor collisions with the moon's surface. It can be shown that a meteor of the right size hitting the moon at the right impact velocity would be sufficient to blast fragments off the moon. Some of those fragments could be captured by the earth's gravitational field [68]. Those who argue this position have to face the difficulty that the chemistry of the Apollo samples does not seem to match that required for the production of tektites.

The origin of tektites is a fascinating topic for study, because those who are concerned with the issue are cleanly and sharply divided in their opinions. There are advocates on both sides who are strongly committed to their respective positions. The tektite issue would appear to be a most fascinating and appropriate subject for concentrated study by historians, sociologists, and psychologists of science. The issue is no more than 100 years old, and so its history is long enough to prove interesting and yet short enough so that one could trace back in considerable detail and with comparative ease the historical development of the issue in the scientific literature [335]. The controversy has many characteristics that are common to all scientific controversies: First, important figures in the history of science are associated with it — for example, Charles Darwin. Second, the different camps or sides appeal to different bodies of evidence to support their positions. They also make use of the same body of evidence, interpreting some of it differently. Some kinds of data they dismiss or reject outright; other data they unconsciously or unknowingly reject. Third, the opposing camps are partly a function of scientific or personality differences and partly a function of different field backgrounds. That is, in part, both camps take different stands and place different emphases or interpretations on the same data because they have been trained to understand and to make an appeal to different kinds of explanatory variables and patterns of reasoning. For these and many other reasons it would be both interesting and important to trace out some of the history, some of the opposing arguments, and especially some of the face-saving or hypothesis-saving devices [122, 254] that have been proposed, utilized, and vigorously defended by both sides to get around the difficulties of their respective positions. However, such an exercise could easily occupy a whole study of its own and is deserving of one. Thus, rather than concentrating on the details of the positions, I will attempt instead to see what, if anything, can be learned from studying the issue from a more global level. Basically the concern will be with the gross attitudinal patterns of the various issues and with the shifts in those patterns.

It should be forthrightly acknowledged that these few issues hardly begin to exhaust the many issues connected with the moon and with Apollo. Thus, attitudes could have been sampled before and during the Apollo missions on a host of other important matters, such as: the age of the moon, whether there are substantial amounts of water on the moon, whether there are or have ever been compounds indicative of life, whether the moon has ever had a substantial magnetic field, whether the moon has a substantial solid core, and various hypotheses regarding the moon's bulk and detailed chemical composition.[2] However, it is not the purpose of this book to examine all of the issues that could be explored. Rather the purpose is to examine some of the important issues so that one of the main theses can be illustrated: that the formation of beliefs with respect to physical issues can be construed as a social science problem. For this purpose, it is not necessary to study every conceivable issue. All that is necessary is to show, for those few issues that are studied, that the social sciences have a valuable and unique contribution to make to the treatment of the issues beyond the usual ways that scientific issues are currently handled in professional journals and meetings. The usual ways of reporting leave out, if they do not systematically repress, all of the issues that are of most interest to the social analyst of science.

Figure 5 represents a plot of the mean sample responses of Table 11. Figure 5 and Table 11 contain the same kind of information, with one important difference. Table 11 looks for and reports significant statistical differences between the mean sample responses of the *columns* of Table 11, whereas Figure 5 looks for and reports significant statistical differences between the mean sample responses of the *rows* of Table 11. An inspection of Figure 5 should help to clarify the differences between these two modes of presentation.

Figure 5 shows, for each time period taken separately, the relative positions of the various hypotheses under each issue. For each time scale, Figure 5 shows (at the extreme right of the figure) whether the differences among the hypotheses are statistically significant. Thus, for example, the difference between the hot and the cold moon positions on the before temperature history time scale is statistically significant at 0.01 level. On the other hand, the difference between the meteorite and lava positions is not significant on the before mascons time scale.

Table 11

Attitudes of the Scientists Toward Various Scientific Issues Over Time

Scientific Issue	Time of Sample Opinion Poll						
Interview Round	III		III		IV		
Point in Time	Before Apollo 11		Now		Later		
Apollo Missions Covered	10 and before		11, 12, 14		14, 15		
	Mean	Variance	Mean	Variance	Mean	Variance	
ORIGIN OF THE MOON							
(1) *Fission hypothesis*: The moon was formed by fission from the earth.	5.00	2.40	5.68	2.57	5.56	2.55	S
		α = 0.01					
(2) *Double planet hypothesis*: The moon and the earth were formed independently of one another but in the same neighborhood.	4.07	2.27	4.24	3.68	3.73	3.05	NS
(3) *Capture hypothesis*: The moon was formed elsewhere in the solar system and subsequently captured by the earth	4.78	2.12	4.58	3.65	4.75	3.08	NS
(4) *Condensation hypothesis*: The moon was formed by condensation from a hot silicate atmosphere of the primitive earth	4.36	2.03	4.02	3.37	3.93	2.31	NS
(5) *Accretion hypothesis*: The moon was formed by accretion from planetesimals of primordial Type I carbonaceous chondrites	3.98	2.97	4.65	3.98	4.39	3.49	S
		α = 0.02					

Table 11 (continued)

		Time of Sample Opinion Poll					
Interview Round		III		III		IV	
Point in Time		Before Apollo 11		Now		Later	
Apollo Missions Covered		10 and before		11, 12, 14		14, 15	
Scientific Issue		Mean	Variance	Mean	Variance	Mean	Variance	

TEMPERATURE HISTORY

Scientific Issue	Mean	Variance	Mean	Variance	Mean	Variance	
(1) *Cold moon*: The moon has never undergone substantial heating or melting to any considerable depth	4.87	4.46	5.59	4.39	5.56	3.30	NS
(2) *Hot moon*: The moon has undergone extensive heating and thorough differentiation to a considerable depth	3.36	3.94	2.56	3.80	2.39	2.49	S

between 3.36 and 2.56: $\alpha = 0.05$; between 2.56 and 2.39: $\alpha = 0.01$

MASCONS

(1) *Meteorites*: Mascons are due to large meteorites buried beneath the lunar surface	3.87	4.20	4.24	4.78	5.14	4.02	S
(2) *Lava*: Mascons are due to large lava plugs which have welled up from the lunar interior	4.39	4.14	4.12	5.61	3.00	3.20	S

between 3.87 and 4.24: $\alpha = 0.01$; between 4.24 and 5.14: $\alpha = 0.01$; between 4.39 and 4.12: $\alpha = 0.001$; between 4.12 and 3.00: $\alpha = 0.001$

TEKTITES

(1) *Lunar*: The origin of tektites is lunar	4.36	3.98	5.20	3.81	5.31	3.87	S
(2) *Terrestrial*: The origin of tektites is terrestrial	3.54	3.40	2.87	3.16	2.90	3.14	S

between 4.36 and 5.20: $\alpha = 0.01$; between 5.20 and 5.31: $\alpha = 0.001$; between 3.54 and 2.87: $\alpha = 0.01$

Note: NS = No significant differences between means; S = Significant difference between means shown by arrows; α = level indicated.

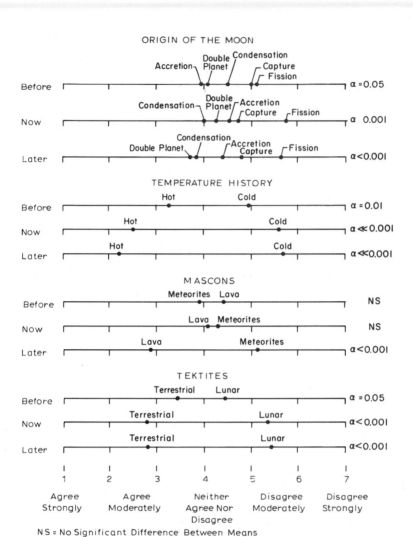

Figure 5. Attitudes of the scientists toward various scientific issues over time.

Table 11, on the other hand, looks for significant differences among the successive time positions for a given hypothesis. Thus, for example, Table 11 shows that the difference between the before and now judgments of the fission hypothesis is significant at the 0.01 level, a relatively high level of statistical significance. Another way to put this is to note that in terms of Figure 5 the difference between the positions of the fission hypothesis on the before time scale and on the now time scale $(5.68 - 5.00 = 0.68)$ is significant at the 0.01 level. For a given time scale, Figure 5 asks if there are significant differences between the positions of the various hypotheses. Table 11 asks if there are significant differences among the successive time positions of a given hypothesis — for example, whether the agreement or disagreement with respect to specific hypothesis is increasing or decreasing.

A more detailed examination of Figure 5 reveals a number of important points about the general nature of the response patterns associated with each issue. The most significant point would appear to be the fact that *for each issue, the spread or the statistical differences between the various competing hypotheses continually increases over time.* In other words, the differentiation between hypotheses becomes sharper over time; the preferred hypotheses move increasingly to the left; the less preferred hypotheses fall by the wayside to the right.

To discuss further points about the nature of the patterns in Figure 5, it is necessary to inspect some of the specific issues. Consider, for example, the hypotheses pertaining to the moon's origin. Fission starts out (Before) somewhat improbable and becomes increasingly more improbable over time. However, in the beginning, the spread among the various hypotheses, although distinct, is not severe. Further, if we ignore the position of the fission hypothesis on the Now time scale, the distinctions among four of the five hypotheses is still minor. In the words of the respondents, all of the hypotheses save possibly that of fission were "still very much in the running at this time." Although the Apollo 11 and 12 data (the Now time period) was, in the words of the respondents, "beginning to set some severe physical constraints on all of the theories, it was not possible to rule out any of them completely." This judgment was still uttered by the time of the third round (Later); however, by then it was voiced considerably less frequently and with much less strength. You will recall from the discussion in Chapter Three that the hypotheses for lunar origin generally fall into three broad classes: (1) fission, (2) capture, and (3) simultaneous formation. Accretion, condensation, and double planet are all variants of simultaneous formation hypotheses. The three forms of simultaneous formation hypotheses jockey among themselves for position on each of the time periods. By the time of the last round (Later), the double planet and condensation hypotheses have become virtually indistinguishable and have clearly separated

155

themselves out as a group from the other hypotheses. However, if general consensus is any criterion for the acceptance or rejection of a scientific hypothesis, then the fission hypothesis was the first to fall by the wayside as indicated by a statistically significant shift in group judgments between the Before and Now time periods.

The pattern under the mascon issue is particularly interesting. It shows that not all of the patterns are of the same kind. The development of the responses with respect to this particular issue show a crossover effect, a reversal between the initial and the final judgments of strongest agreement and disagreement. Finally, the tektite issue pattern is interesting because it is the only issue for which there is a significant difference between both hypotheses on the Before and Now judgments.

Figure 5 and this whole mode of analysis raise more questions than they answer at this time. What is the range of actual developmental patterns? What is the range of possible patterns? Under which conditions do crossovers or reversals normally occur? How frequent are they? At which point do the curves begin to flatten out (do their slopes approach zero)? Just exactly what can be inferred about the course of scientific development or progress from the properties of such patterns or curves? And finally, what significance do the various patterns have for the philosophy, sociology, and psychology of science? These are only a few of the many questions that such an analysis raises. And indeed, this is precisely one of the main purposes of the analysis and discussion — to raise new questions regarding the development of attitudes with respect to scientific issues and to show that these questions can be pursued. Before one can answer all the questions one would like to have answered on a phenomenon, one first has to show that there is a viable way of posing the questions, that there is a way of conceptualizing the phenomenon of interest. This has been the major purpose of this section — to show that one can begin to demonstrate and detect developmental patterns and changes in the course of scientific issues.

In this regard, the lunar hypothesis for the origin of tektites gives an indication of the kinds of signs that are important to look for with respect to the patterns. A statistically significant shift between the positions of the lunar hypothesis on the Before and Now time scales can be explained as a clear and direct indication of the decisiveness with which the majority of the scientists regarded the Apollo 11 and 12 data. The verbal protocols collected in conjunction with the quantitative scale responses indicate a firm rejection of the lunar hypothesis. Time and time again I heard the statement, "The chemistry of Apollo 11 and 12 samples just doesn't match that of tektite glass; the tektite issue is dead." This strong shift in attitude is most interesting, because not too many years before the flight of Apollo 11 an important conference was held on the subject of tektites. At the end of that conference a show of hands was asked of the audience. At that time, the lunar and terrestrial hypotheses each received roughly 50 percent of the votes of the respondents who were present. This may help to account for the fact that the terrestrial and lunar hypothesis equally bracket the 4 or neutral position on the Before time scale. By the time of Apollo 12 (the Now time scale), the number favorably inclined toward the lunar hypothesis had fallen off dramatically. Conversely, those few scientists who continued to act as strong advocates for the lunar hypothesis fell further and further out on the distribution of responses. In the language of statistics, they fell significantly many standard deviations away from the mean of the distribution. In the language of social psychology, their position became deviant. This, of course, does not prove that one side has been finally proved right on the issue and the other side wrong. It also does not say that the advocates for the minority position should give in and cease making noise altogether. Although it is true that many of the respondents felt this way, there were also those who felt that no issue was ever completely settled and that therefore one always needed advocates on the other side. However, even here, most felt that the case was pretty hopeless and perhaps ought to be closed: "The tektite issue is dead; there're more important things to get on with; it never should have been blown up and accorded that much importance in the first place." In short, the point of this kind of analysis is not to prove that one side or the other is right, but rather to demonstrate that one can begin to chart the course and development of scientific opinion. The long term hope is that this way of viewing and presenting scientific issues will be of interest and will help the participants to gain a novel perspective on themselves and the issues.

Before concluding this section, I want to emphasize that Table 11 and Figure 5 are not the only ways to conceptualize and measure the phenomenon of the formation and shift of scientific attitudes. Although there are, in principle, a large number of additional conceptualizations that could be explored, there is one approach that seems especially promising for future investigations.

The approach I have in mind considers the shift in scientific attitudes as an *information transfer problem*.

It can be shown [437] that if we have a system of mutually exclusive events (or in our case, hypotheses), $E_1,...,E_n$, with prior probabilities $p_1,...,p_n$ then the expected information of the message which transforms the prior probabilities to the posterior probabilities $q_1,...,q_n$ is given by the following expression:

$$I(q:p) = \sum_{i=1}^{n} q_i [\log_2(q_i/p_i)] .$$

That is, if we consider the attitudinal judgments on the various hypotheses as judgments of their respective probabilities, and if we also consider the various Apollo missions as evidence messages that act to revise these probabilities, then we can compute the information transferred between time periods as a result of the Apollo missions.[3] Table 12 presents the results of such an analysis.

The amounts of information transferred in Table 12 are extremely small. For example, consider the hypothetical case where both hypotheses start out equally uncertain: $p_1 = p_2 = 0.5$; if one of the hypotheses goes to certainty, say $q_1 = 1.0$, then $I = 1$. If one takes a more extreme hypothetical case where $p_1 = 0.1$, $p_2 = 0.9$, $q_1 = 1.0$, $q_2 = 0$, then $I = 3.32$; that is, I is greatest for a reversal as in the case of mascons (see Table 12). Measured against these hypothetical cases or standards, the information transferred by the Apollo missions was extremely small. This was most readily apparent on round III. The respondents were required to give two simultaneous responses side by side, namely how they felt on the issues "Before Apollo 11" and "Now." The most frequently heard response as the respondents checked the scales was, "Huh, I guess this shows that I haven't changed much. I felt this way before Apollo 11 and I still do now. I don't see much reason to change. It takes time to sift the evidence." In this special sense, the respondents recognized the slowness of change in their own attitudes.

Obviously, whether a change is big or small depends on how we look at it. Indeed, this is the whole point. If we look at the shift in judgments as the difference between the means of two distributions, then the change in attitudes due to the Apollo missions is significant in the statistical sense (Figure 5). If we look at the matter in information transfer terms, the change or difference between time periods is exceedingly small. But then again the

158

Table 12.

Information Transferred Between Time Periods As a Result of the Apollo Missions

Scientific issue	Interview round Point in time Apollo missions covered	II Before 10 and before	III Now 11, 12, 14	IV Later 14, 15	Information transferred in bits		
					I (Before: Now)	I (Now: Later)	I (Before: Later)
Origin of the moon		Normalized probabilities					
Fission		0.204	0.136	0.136			
Double planet		0.315	0.294	0.328	0.0307	0.00457	0.0249
Capture		0.213	0.245	0.221			
Condensation		0.268	0.325	0.315			
Temperature history							
Cold		0.353	0.238	0.216	0.0447	0.00184	0.0637
Hot		0.647	0.762	0.784			
Mascons							
Meteorites		0.544	0.481	0.300	0.0114	0.0980	0.175
Lava		0.456	0.519	0.700			
Tektites							
Lunar		0.418	0.284	0.274	0.056	0.00030	0.0642
Terrestrial		0.582	0.716	0.726			

question is, "Small compared to what? What is the standard?"[4] Obviously a study of other cases, historical and contemporary is needed, so that we can begin to form an idea of what constitutes a criterion for distinguishing large from small cases of information transfer. It is interesting to note that the information transfer between Before and Later, for a more global and more difficult to settle issue like the origin of the moon, is nearly one-fourth that for a more specific and apparently more easily settled issue like the temperature history of the moon, and roughly one-seventh of that associated with the issue of mascons. This raises the interesting possibility that one might be able to characterize the difficulty of some issues in terms of the amount and rate of change in their associated information measures. Is it a characteristic of issues that are more difficult to settle that the information transfer associated with them is low? Is this a useful way to view issues? Is it anything more than a trivial tautology? The questions certainly deserve more investigation.

Finally, the various modes of analysis performed in this section raise one of the most fascinating questions of all: How should the judgments of the individual respondents be weighted? All of the analyses performed in this section assume an equal weighting, but obviously not all scientists are equal. Certainly the sample did not feel they were. The judgments of some scientists always carry more weight than those of others. However, once this was agreed to, there was far less agreement by the sample with respect to whose judgments should get top weighting. No matter how one views it, the understanding of the moon is a problem for the social sciences. All of the sciences face the problem of how the judgments of individual scientists should be combined into the collective judgment of the science. The analyses of this section bring this problem directly to the forefront.

Perceived Commitment to Specific Issues

Max Planck wrote: "A new scientific truth does not triumph by convincing its opponents and making them see the light, but rather because its opponents eventually die, and a new generation grows up that is familiar with it" [341, pp. 33-34]. In the last chapter the psychological portraits of some specific scientists were examined in detail without reference to any particular scientific issues with which they might be identified. Thus, for example, we examined how biased or impartial some scientists were perceived as being as a general personality trait. However, ultimately we are interested in how the psychology of science bears on scientific issues, and so it is appropriate to examine how committed these scientists were perceived as being in relation to some specific issues, not just in the abstract. Are the portraits of these scientists in relation to some specific issues the same as the portraits obtained by looking at these scientists in general? Is there a differential perception of commitment; are some scientists perceived as being more committed to some issues than other scientists? If so, can one measure this commitment? Is it possible to measure the degree of a scientist's commitment to an issue? And finally, is it possible to measure the shift in commitment? Which scientists are perceived as more enduring in their commitments than others?

Ideally I would like to show that scientists have strong prior commitments or beliefs, which by now should come as no big surprise. But I would also like to take the next and more crucial step of demonstrating in detail how these strong beliefs affect a scientist's assimilation and interpretation of subsequent data and arguments. In principle this could be done for the particular set of scientists and issues which are being dealt with here. Some of the issues and the positions with respect to those issues are so clear-cut and so well articulated that one could undertake a direct study and demonstration of the impact of prior belief-systems [376] on scientific data and arguments. Indeed, much of the interview data was collected with this purpose in mind, and so this could be done. However much this exercise is called for, such an exercise unfortunately cannot be carried out at this time. Such a task would not only consume substantial space and time, but most important of all, it would necessitate the breaking of strict promises of confidentiality. It is necessary to preserve the anonymity of the respondents for the immediate future. For this basic reason we shall have to content ourselves with how much we can learn from some general observations on the perceived commitment of the specific scientists we have been studying — $S_I^{\,1}$, $S_I^{\,2}$, $S_I^{\,3}$, S_{II}, S_{III}, and Y. Neither the specific scientists nor the specific issues to which these scientists were perceived as being committed can be identified at this time. Some of these scientists are so strongly associated with particular issues that merely to identify the issues would be to identify the scientists. Thus in Figure 6, the issues which underlie scale 1 must remain unspecified.

However, I must also say that, although I would like to engage

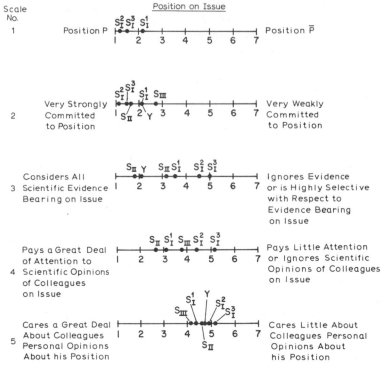

Figure 6. Specific commitments of specific scientists.

in a detailed examination of the effect of prior beliefs, a more global analysis of perceived commitments is not without its benefits. The results of such an analysis are but further testimony to the fact that the respondents knew and explicitly acknowledged that scientists had strong prior commitments that affected the detailed doing and interpretation of science. This is valuable knowledge if we are finally to crack the myth that scientists are dispassionate creatures who do science with open minds, untainted by strong prior beliefs about what they expect and even hope to find. The documentation and persistence of perceived commitment is an important first step in its own right in the eventual examination of the detailed effects of commitment.

Figure 6 presents the five scales with respect to which the commitment of the five specific scientists S_I^1, S_I^2, S_I^3, S_{II}, and S_{III} were rated by the sample. Again, Y represents the response category Yourself. Each of the plotted points — S_I^1 and so on — represents the average of the sample responses.

162

On every scientific issue, especially unsettled issues, there is probably a large if not infinite number of contending positions or stands that one could adopt. For some scientists, the different positions will take the form of different detailed theories or working hypotheses; for others, the different positions will be as broad as opposing global stances, world views, or scientific paradigms [247, 248]. Further, some of these different positions will undoubtedly overlap and thus be in agreement with one another; others will undoubtedly be in sharp disagreement — for example, in the form of contrary positions.

In the measurement of perceived commitment, the positions for each issue were represented as contraries in the form of opposing ends of a single scale. From the preceding interview rounds it had been determined that the positions with which certain of the scientists were identified were not merely in mild or weak disagreement with one another, but in sharp opposition. In brief, some of the positions and scientists were locked in strong adversary proceedings. Thus, each respondent was asked to rate the position of each of the specific scientists, including himself, with respect to a single pair of opposing positions for a particular scientific issue. For each of the specific scientists, each respondent indicated which of the two positions or poles tended to best represent that scientist's position on the issue. If a respondent felt that position P best represented a scientist's position, he would check a number near the 1 end of the scale; if he felt $\bar{P}$ (the contrary of P) was more representative, he checked a number near the 7 end of the scale; if a respondent felt that neither pole represented a scientist's position, he was instructed to check the middle position, 4.

This was the procedure that was followed with the three scientists S_I^1, S_I^2, and S_I^3. From the initial interviews a set of contrary positions could be easily formulated, prior to the exercise, for each of the scientists S_I^1, S_I^2, and S_I^3. These scientists had already been identified by the sample with particular issues. As a check on whether the respondents still agreed with their earlier characterizations of the polar positions, prior to the scaling exercise the respondents were asked whether the polar positions were a fair and accurate representation of the dispute or issue in which these scientists were embroiled. Only in a very few instances did the respondents disagree with the polar positions that were presented. Where they did, they were asked either to modify the positions displayed or to substitute poles of their own liking. Only in a very few instances was a respondent unable to do this. Thus, with

163

these scientists essentially all a respondent had to do was to place a check mark toward one of the two preassigned poles. With the remaining scientists — S_{II}, S_{III}, and Y — the respondents had to do considerably more. First they had to indicate whether they were aware of or could formulate a set of polar positions in which these scientists were embroiled. They then had to indicate how much they felt a particular scientist leaned toward these poles.

Although the results of Figure 6 are not in exact accord with those of the previous chapter, the agreement is more than good enough for the same general pattern to emerge: S_I^1, S_I^2, and S_I^3 are not only perceived as being more committed or biased in general, but they are also perceived as being more committed on specific issues. Furthermore, although the order is not exactly the same as before, there is a definite order to their degree of perceived commitment. Because the scales of Figure 6 are largely self-explanatory, a detailed discussion of all but the first is omitted.

Scale 1 is a composite of many issues. The polar positions P and $\bar{P}$ are different for each of the three scientists whose locations on scale 1 are shown. Thus, scale 1 is meant to show how close each of the three scientists was perceived as being to one of the polar positions; each set of positions is different for each of the scientists.

The most significant thing about scale 1 is what is not shown. The locations of S_I^1, S_I^2, and S_I^3 are shown on scale 1 because (1) a set of polar positions could be easily formulated for these scientists, and (2) their location with respect to these positions could be easily scaled by the respondents. The remaining scientists S_{II}, S_{III}, and Y are not shown because neither of these two requirements could be fulfilled; 32 out of the 42 scientists who responded to the exercise, or 76 percent of the respondents, were unable to specify any set of polar positions in which S_{II} and S_{III} were embroiled. Likewise, 17 out of 42, or 41 percent of the respondents, were unable to specify any set of polar positions in which they were embroiled. The significance of these results is twofold: (1) Once again the differences between different types of scientists can be observed; type I scientists are not only more likely to be committed to a particular position, but they are also more likely to be embroiled in an issue or controversy and to be strongly and easily identified with it. And (2) once again the sensitivity of different kinds of measuring procedures can be witnessed — for example, the sensitivity of the perceptions of the sample respon-

164

dents to differences that other exercises fail to pick up between the types of scientists. For example, covariance analyses of the positions of the scientists across all the time periods on the issues of the last section fail to pick up any statistically significant differences between the scientists grouped according to their type (see the 6 types of Chapter Three). That is, although there certainly are marked differences between the positions of some of the individual scientists on the various issues of the last section, there are no significant differences between their positions when they are analyzed in terms of their average group score differences. As groups, the scientists take essentially the same stands on the issues and cannot be easily differentiated from one another. As individuals, particularly as they are perceived by their peers, the situation is very different; they can be readily distinguished from one another. This conclusion parallels those of the previous chapter.

A number of important implications follow from these results. Given the current ways of defining the types and the current limitations of measurement or classification of a scientist with respect to the typology, one cannot predict from a knowledge of a scientist's type alone what his position on a specific issue is likely to be. Whether a scientist is more a speculative theorist or more a data-bound experimentalist is by itself not sufficient to tell us whether he is more likely to agree or to disagree with a particular issue. However, what one can do is to make a very strong prediction about the types of scientists, for example, the type I scientists are more likely to become committed to positions and to take strong stands, even if one cannot predict the exact direction and amount of that commitment on a particular issue. However, even within the class of type Is, you cannot predict the individuals' positions from the group's position, because the group or average position blurs crucial distinctions between the individuals. In other words, given the kind of analysis one is able to perform at this stage in the development of the social sciences, it is possible to begin to make important gross attitudinal distinctions between different types or kinds of scientists, but from this alone one cannot make predictions about specific stands on specific issues.[5]

This is not to say that other analyses using other variables — for example, different classification schemes based on the institutional affiliations of the scientists — would or would not be able to achieve more detailed predictions. However, it is not even clear that this should be the goal, although it would be desirable, at this stage in development. Rather, this discussion is intended to point out the limitations of what can be currently accomplished. Given the current limitations of social science research, just being able to establish consistent significant differences between various types of scientists may be adequate, particularly if this helps to stimulate further research in an area that has not been overly studied. In the early historical development of a field, which is where the psychology of science still is, it is just as important to point out the limitations of a method of analysis as it is to point out its strengths and accomplishments.

Because of the large number of additional items that were of interest to measure on the last round of interviews, and because it was mainly the three scientists $S_I{}^1$, $S_I{}^2$, and $S_I{}^3$ who were perceived as committed to specific issues, the three scientists S_{II}, S_{III}, and Y plus some of the scales were dropped on the last round. Figure 7 shows the differences or shift, between the perceived

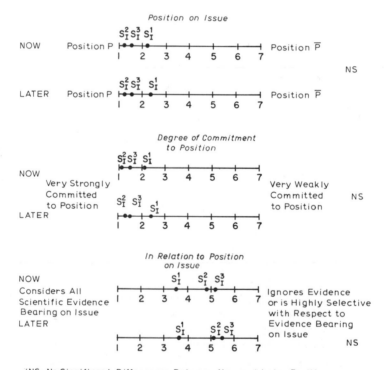

Figure 7. Change of specific commitments of some specific scientists.

166

commitment of S_I^1, S_I^2, and S_I^3 to specific positions on certain issues, from interview round III (Now) to round IV (Later). *The significant thing about Figure 7 is that there are no significant differences.* The perceived numerical shift is virtually zero for the three specific scientists. We have a numerical picture of the perceived commitment of some specific scientists, and we also have a picture of how enduring that perceived commitment is. In the words of a number of the respondents: "They just don't change, do they? But then, perhaps if I were honest with myself I'd say I haven't changed much either and again I'm not so sure that it is always bad for science for scientists not to change easily, although it can be extremely dangerous and irritating at times."

Jungian-Based Approach to Measurement of Different Kinds of Scientists

It would be helpful if the existence of something as important as the notion of fundamentally differing kinds of scientists were based on something more substantial than the mere fact of significant empirical differences between some specific representatives and the mere suggestion of a crude typology as first proposed by some of the respondents. It would be helpful if a typology of different kinds of scientists could be derived from a more generic typology which itself possessed some kind of theoretical justification. This section deals with precisely this problem.

If the construction of a typology of different personality types were merely a problem for empirical psychology, then any one of a number of typologies and different approaches to the measurement of personality already in the psychological literature could be used. However, if one believes that the construction of a typology is as much a conceptual problem which requires a broad philosophical perspective as it is an empirical and a technical problem, then this narrows the choice of available approaches considerably. For reasons that will become apparent, the approach adopted here was that of C.G. Jung.

The reader who is interested in learning in detail about the Jungian system is referred to the literature [213-216]. The following discussion is intended mainly to convey the spirit of the system so that the reader can appreciate the underlying basis from which the new system of different scientific types was derived.

The Jungian personality typology is characterized by four major modes or psychological functions. Two of the modes pertain to the dominant psychological functions that an individual uses to *perceive* (sense) the objects of the world, while the other two modes pertain to the dominant psychological functions that the individual uses to *evaluate* (judge) the objects of his perception. Because the functions for perception are presumed to be independent of the functions for evaluation, four perception-evaluation combinations result. The alternate modes of perception are *sensation* and *intuition*. The alternate modes for evaluation are *thinking* and *feeling*. In most individuals, a preference for one mode of perceiving and one mode of evaluation is characteristically developed. The alternate modes remain, as a result, underdeveloped or unconscious. When a particular mode or function becomes especially dominant or developed to an extreme degree within an individual, then one speaks of a *psychological type*.

A preference for sensation refers to that type of individual who relies primarily on data received by his senses in order to perceive the objects of the world. The reality of the sensation type, in other words, is typified by and grounded in sensory processes, objective, hard facts, and attention to detail. Intuition refers to the mode of perceiving objects as possibilities. Whereas sensation perceives objects as they are, in isolation, and in detail, intuition perceives objects as they might be and in totality, as a gestalt. Notice that although these modes are in conflict, neither one is superior or more fundamental than the other. If it is the virtue of sensation types that they are guided by the facts and are careful not to extrapolate beyond them, then it is the virtue of intuition types that they see through the facts and do extrapolate beyond them. Whereas the sensation type may be too data-bound (he tends to go on collecting data forever because he is afraid to risk a generalization that goes beyond the available data), the intuition type may be too data-free; he may spin out a hypothetical conclusion a minute that is based on no data at all.

William T. Morris's paper on "Intuition and Relevance" [319] vividly demonstrates the importance of both of these types within the field of management. Every corporation needs both kinds of managers. A management composed solely of sensation types runs the risk of being too limited and bound by the current set of facts; it runs the risk of being unable to envision future possibilities completely contrary to the current state of affairs; it has a short planning horizon. A management composed solely of intuition

168

types runs the risk of always living in the future and thus never paying proper attention to the present.

The alternate modes for evaluation are thinking and feeling. A preference for thinking is exhibited by the type of individual who relies primarily on cognitive processes for subsuming reality. His evaluations tend to run along the lines of abstract true/false judgments and to be based on impersonal, formal systems of reasoning. A preference for feeling is exhibited by the type of individual who relies primarily on affective processes. His evaluations tend to run along personalistic lines of good/bad, pleasant/unpleasant, and like/dislike. He tends not only to make but to like making moral judgments. Thinking systematizes; it builds systems; it analyzes; it defines and makes precise distinctions so that men can be clear and rational about what concerns them; it abstracts from reality to find a set of universal principles or laws that remove the inevitable differences between all concrete individual things. Feeling, on the other hand, individuates; it draws out and heightens the individual differences between all things; it often does this by provocation; feeling often offends; it satirizes all rules, definitions, and systems; it pokes deliberate fun at precision and thinking; it makes all things unique, individual. As in the case of the two perceptual modes, the two evaluative modes also tend to be mutually exclusive. An individual characteristically tends to prefer one type more than the other. Combining dominant modes in all permissible ways results in four major psychological orientations towards the world: thinking-sensation, thinking-intuition, feeling-sensation, and feeling-intuition. Where one combination especially dominates an individual's whole outlook, we may again speak of a psychological type that is typified by that particular combination.

If only because it caricatures, the notion of types is helpful in making headway in the elusive problem of personality delineation. In contrast to other typologies, the Jungian system constantly stresses that the pure types are theoretical constructs only. It is the blends and contrasts between these pure types that make up the actual living personality. The Jungian typology also constantly stresses that no one mode of perceiving reality is ultimately more basic than or superior to another. Indeed, it is the belief in the ultimate superiority of one function or another that types an individual. We might say that it is a particular type who believes in the actual existence of types.[6] As one of my mentors, Thomas A. Cowan, has put it, "There are two kinds of people: 'those who think there are two kinds of people' and those who don't."

169

Cowan's aphorism is a perpetual challenge (and a satirical one) to those who believe in the permanent existence of a finite number of types sufficient to describe the entire range of humanity.

For further details and supporting evidence on the existence or meaningfulness of these types, the reader is referred to the work of Allport and Vernon [21], Ginzberg [148], Gundlach and Gerum [162], Murray [325], Spranger [417], and Thurstone [440]. The reader is especially referred to Mogar [317] for an interesting discussion of the Jungian system. Finally, see Ackoff [16] for a stimulating discussion of how the Jungian types can be made operationally precise to meet the demands of the formalist (thinking) and the empiricist (sensation). For a popular exposition of Jungian psychology, see [217].

Short character sketches or portraits, based on my understanding and interpretation of Jung plus the results of the previous chapters on the defining characteristics of the different types of scientists, were constructed for four very different kinds of scientists.[7] The four character sketches represent my translation of Jung's four personality types into the realm of science, the particular form as I see it that these types can take in the field of science. In this regard, it could be said that there is no standard translation. I do not pretend that every investigator would utilize Jung or translate him as I have. The particular character sketches should in no way be regarded as unique representations.

In constructing the portraits, every effort was made to balance them, to have them all be of equal attractiveness, of equal emotional strength, and even to have the sketches be of equal length. Everything was done to avoid making any one of them more desirable than the rest. The extent to which this was not achieved is a testimony to my psychology, the fact that I have an initial and strong preference for one of the types which colors my whole outlook. I have to acknowledge that my own psychology was present throughout the entire study, as it could not help but be. As Jung himself put it with great candor and honesty:

> Nobody is absolutely right in psychological matters. Never forget that in psychology the *means* by which you judge and observe the psyche is the *psyche* itself. Have you ever heard of a hammer beating itself? In psychology the observer is the observed. The psyche is not only the *object* but also the *subject* of our science. So you see, it is a vicious circle and we have to be very modest. The best we can expect in psychology is that everybody puts his cards on the table and admits: "I handle things in such and such a way, and this is how I see them." Then we can compare notes [214, p. 142].

I decided to present the Jungian types to the scientists in form of short character sketches for a very definite reason. I fel that enough had been done in previous rounds, to get the respondents' conceptions and reactions to the various types of scientists along simple isolated adjective scales, to continue obtaining this kind of data. Enough previous research has been done to show that people do not conceive of other people in this manner [26, 442]. Even when we present seemingly isolated scales, people form whole images of the things they are scaling. Thus it was felt desirable, at some point in the study, to give the respondents some whole portraits to react to. Finally, it should be noted that the sketches contain strong positive and negative statements. This, too, was done for a very definite reason. Each of the characters or types is not content merely to sing his own praises and virtues loudly — he also feels obliged to get in a strong dig at others. This is one of the defining characteristics of a type; he has little tolerance for other modes of perceiving and evaluating reality. Each of the four character sketches is given here, with the instructions given to the respondents.

Below you will find descriptions of four very different kinds of scientists. I would like you to read each description carefully, and then only after you have read each description, indicate the degree to which each description represents you.

Type A: The Hard Experimentalist. Type A is the kind of scientist who first and foremost regards himself as a *Hard Experimentalist.* He takes extreme pride in his carefully designed and detailed experimental work. In general, he prefers hard data gathering to abstract theorizing, intuitive synthesizing, or humanistic concerns. He feels that one really doesn't understand something until he has collected some hard data on it. He feels that abstract theorists have a tendency to get lost in their abstractions for their own sake and hence to mistake them for reality, that intuitive synthesizers have a tendency to engage in unwarranted extrapolation beyond the data at hand, and that humanistic scientists have a tendency to become prone to gushy moralizing. His attitudes toward theorizing and speculation are modest. He feels that theorizing and speculation are only warranted when the data are available that clearly support such activities. He is quick to master complicated and sophisticated experimental techniques. He prefers to work on manageable, well-defined problems for which there are available standard, well-developed experimental methods of investigation. He tends to be technique-oriented rather than problem-oriented. In sum his approach to science is best described as Empirical-Inductive rather than Theoretical-Deductive.

Type B: The Abstract Theorizer. Type B is the kind of scientist who first and foremost regards himself as an *Abstract Theorizer.* He takes

171

de in his ability to construct formal, analytical models of physical phenomena. In general, he prefers building ab-
etical models to experimental data gathering. He feels that loesn't understand something until he has built a general t. He feels that hard data gatherers have a tendency to ingrossed in collecting data for its own sake that they never to putting it all together in some systematic conceptual so feels that intuitive synthesizers and humanistic scientists both have a tendency to be extremely fuzzy in their thinking. His attitude is that the construction and investigation of formal models and theories produces the best analysis and understanding of scientific prob-lems. In this sense he is extremely critical of speculation that is not tied down and checked by formal reasoning. He is quick to master compli-cated and sophisticated analytical techniques. He prefers to work on manageable, well-defined problems for which there are available stand-ard, well-developed analytical methods of investigation. He tends to be technique-oriented rather than problem-oriented. In sum, his approach to science is best described as Theoretical-Deductive rather than Empiri-cal-Inductive.

Type C: The Intuitive Synthesizer. Type C is the kind of scientist who first and foremost regards himself as an *Intuitive Synthesizer.* He takes extreme pride in his ability to synthesize and intuit the meaning of a wide variety of experimental and theoretical facts and ideas. In general, he prefers extrapolation from and speculation on existing data to gathering data of his own. He feels that one doesn't really under-stand something until he has developed a deep intuitive insight into the basic meaning of that something. He feels that hard data gatherers have a tendency to go on collecting data forever because they lack the basic intellectual or emotional fortitude that would permit them to extra-polate beyond their always limited sets of data. He also feels that abstract theorizers are equally limited, for example, their overly formal-istic ways of conceptualizing phenomena prevent them from appreciat-ing characterization of problems that are not easily, if ever, susceptible to formalization. (Humanistic Scientists he tends to dismiss as irrele-vant.) His general attitude is that intuition and a global approach pro-duces the best ultimate understanding of scientific problems. This, of course, is a reflection of the fact that his understanding of physical laws and processes is more intuitive than it is formal or even precise. He is quick to formulate and take in broad, sweeping views of problems. He is quick to generate a large number of interesting hypotheses about any problem. He has a high tolerance and even preference for ill-structured problems, the problems that others tend to shun. He tends to be more problem-oriented than technique-oriented. In sum, his approach to science is best described as Intuitive-Synthetic rather than as Theoreti-cal-Deductive or Empirical-Inductive.

Type D: The Humanistic Scientist. Type D is the kind of scientist who first and foremost regards himself as a *Humanistic Scientist.* He takes extreme pride in his ability to perceive the political and moral implications of scientific work and discoveries. In general, he is more concerned with being able to predict the desirable versus undesirable

172

consequences of scientific products than he is concerned with the details of scientific method that generate the end products. He feels that scientists have been extremely derelict in contributing to the general moral and political understanding of their discoveries. He feels that hard experimentalists, abstract theorizers, and even intuitive synthesizers take too narrow and restrictive an attitude toward science. They are all too preoccupied with the detailed tools and techniques of scientific method, than they are with evaluating the overall consequences of their end-products. He feels that they are much too insensitive towards the moral and human elements in science. For too long, he feels that scientists have kidded themselves that they could study physical phenomena in a completely detached and objective way. He feels it is high time for scientists to realize that their subjective feelings and emotions deeply affect their so-called objective studies and descriptions of nature. This type of scientist also tends to be more problem-oriented than technique-oriented. In sum, his approach to science is best described as Personally Involved and Evaluative rather than Detached-Empirical-Analytic.

While they were reading each portrait, the respondents were explicitly encouraged to comment freely on how the portraits struck them. Immediately after and only after reading all the portraits, the respondents were asked to indicate on a seven point Likert scale the degree to which each portrait represented them. On the scale, 1 indicated complete and accurate representation, 7 indicated complete misrepresentation or inaccurate representation, and 4 indicated that a portrait neither represented or misrepresented an individual. Thus, a respondent if he so desired could check all 1s, 4s, or 7s, or any number in between, on all the scales. There was no need for any or all the portraits to represent an individual. He could accept or reject them all with reference to himself or with respect to others as well.

Table 13 and Figure 8 present the results of the Jungian analysis. Table 13 presents the means of the sample responses to each of

Table 13.

Jungian Types

Jungian types	Type	Mean	Standard deviation
Hard experimentalist (sensation)	A	3.90	1.74
Abstract theorizer (thinking)	B	4.85	1.59
Intuitive synthesizer (intuition)	C	2.85	1.48
Humanistic scientist (feeling)	D	4.77	1.66

173

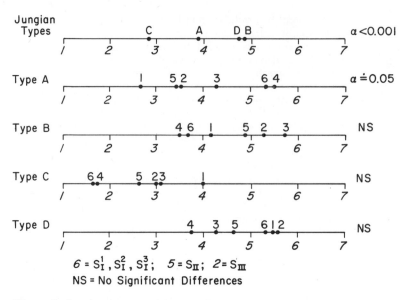

Figure 8. Jungian types and types of scientists.

the Jungian portraits. The first scale of Figure 8 presents a visual plot of the means of Table 13 and shows graphically where the various portraits lie in relationship to one another. The α value to the extreme right of this scale shows that the difference between the responses to the portraits is statistically significant as determined by an analysis of variance.

It is both interesting and significant that the Intuitive Synthesizer was seen as the most representative of all the types. This is significant because it is a commentary on the respondents perceptions of themselves, and also because it is a commentary on the respondents perceptions of the representativeness of the types that the earth and planetary Sciences have tended to attract. This came out in response to two supplementary exercises.

In addition to quantitatively scaling how much each portrait represented himself, each respondent was asked to comment freely on (1) how good or bad each portrait was, and (2) whether, as they read each portrait, the names of any specific scientists who strongly fitted each portrait came quickly to mind. With regard to the first supplementary exercise, nearly all of the respondents felt that the portraits, although caricatures in the sense of being extremes, were nonetheless good in the sense that they captured significant characteristic distinctions among some fundamentally

174

differing kinds of scientists.[8] The most persistent objection raised was to that part of the Hard Experimentalist that described him as "technique-oriented rather than problem-oriented" and as preferring "to work on manageable, well-defined problems for which there are available standard, well-developed experimental methods of investigation." A number of the respondents objected to these descriptions, particularly those who either saw themselves or were seen by others as Hard Experimentalists. A number of the respondents felt that these attributes tended to make the Hard Experimentalist into too much of a technician, that many experimentalists were much broader than this description, that the latter qualities were not absolutely essential or necessary to the characterization of the Experimentalist. And in a sense this is correct. These characteristics are not inherent or necessary qualities of the Experimentalist. However they are necessary characteristics and consistent with the definition of a type. Precisely those characteristics of the Hard Experimentalist that the respondents objected to and wished to replace by others are the very characteristics that are included under the definition of the Intuitive Synthesizer as consistent with the Jungian definitions of sensation and intuition. In other words, the Intuitive Synthesizer possesses exactly those qualities which the Hard Experimentalist does not — and vice versa. To the extent that a respondent's image of himself or of the various types was different, I would expect him to find fault with one or more of the portraits. The more a respondent felt he was a strong blend of all the types, the more I expected him to be dissatisfied with all of the portraits. The fact that the respondents found so little fault with the portraits can be interpreted as a testimony to two things: (1) that the portraits captured some salient characteristics of scientists of which the respondents were well aware, and (2) that the respondents were, with relative ease, able to recognize themselves and their colleagues in the portraits.

With regard to the second supplementary exercise, although nearly all of the respondents felt that the portraits were extremes, they nevertheless had little difficulty in picking out scientists who were examples of those extremes. The clearest cut choices in this regard were those scientists previously designated as S_{III} (for the Hard Experimentalist) and $S_I{}^3$ (for the Intuitive Synthesizer). The choice of these scientists was virtually unanimous; S_{III} and $S_I{}^3$ were seen as the virtual embodiment of the Hard Experimentalist and Intuitive Synthesizer respectively. Far more difficult to name or to agree on were representatives for the Abstract Theorizer and

Humanistic Scientist. Although a few names were mentioned for the Abstract Theorizer, many of the respondents mentioned that there were far fewer outstanding examples of this type within their field of science than in others. The overwhelming majority of the respondents mentioned that their field was predominantly made up of Hard Experimentalists and Intuitive Synthesizers. The dominant response was that it was extremely difficult to be an Abstract Theorizer or a formalist within the field of earth and planetary science, that the problems were just too complex to approach purely in this mode, and perhaps most important of all, that very few scientists were basically good enough to do outstanding work in this mode. One had to be very good to be an Abstract Theorist. It is interesting that S_I^1 was seen as a blend of the Hard Experimentalist and the Intuitive Synthesizer; S_I^2 was seen as a blend of the Abstract Theorizer and the Intuitive Synthesizer.

Of all the types, the hardest for the respondents to associate names with was that of the Humanistic Scientist. The overwhelming majority of the respondents were unable to name anyone within their field who fitted this category. When they did name someone, it was a person like Linus Pauling or Albert Einstein. And in fact this was one of the major motivating points of the whole exercise. From a Jungian point of view, the question is precisely whether all of the types are equally represented within a given field. Theoretically all the types are equally possible within any field or individual, and so it is both interesting and important to note which fields tend to overemphasize which functions.[9] According to Jung's analysis of modern science, we should expect the psychological functions of sensation and thinking to be overemphasized, because they are in modern science's dual emphasis on empirical data and objective reality (sensation) on the one hand and systematic, impersonal method (thinking) on the other hand. Likewise, we should also expect to find intuition and feeling underemphasized because they tend to be perceived as antithetical to the notion of science, or as vague, inherently subjective qualities of thinking.[10] The results for this particular group of scientists are mixed: Intuition is overrepresented contrary to prior expectations, and feeling is underrepresented in line with prior expectations. With regard to the low representation of feeling within the specific science with which we are dealing, the most telling commentary may be that the most frequently heard response to the portrait of the Humanistic Scientist was: "I don't regard this type as a scientist; those who can't do science, or who no longer want to do science, do this." Or, "This type is more politician than scientist." Although this may be true from the perspective of the respondents, it is not true from my perspective. As I indicated in the previous chapter, there are good grounds for questioning whether the role of the scientist and the politician are separate in fact and, even more, whether they ought to be. Again I want to emphasize that not all the respondents felt this way. Some of the scientists identified themselves with the *Humanistic Scientist*. It was rather that, in proportion to the number of scientists (virtually the entire sample) who indicated that they were directly and actively engaged in normal scientific pursuits as indicated by their identification with the first three types, the number of scientists who indicated or gave evidence that they seriously *acted on* their identification with the Humanistic Scientist was extremely small.

If one recalls the earlier definitions of the six types of scientists discussed in Chapter Three, [11] then the bottom four scales of Figure 8 constitute a weak but nevertheless further validation of the six different types of scientists. The bottom four scales show the relative location of the mean scores of the sample broken down according to the six types or groups of scientists into which the sample fell. On each of the Jungian portraits or scales, the relative placement of each of the six different types of scientists is in general accord with what we would expect. Thus, for example, the Hard Experimentalist or Type A portrait was perceived as most representative of those scientists previously designated or classified as type 1s, as it should be. Likewise, the Hard Experimentalist portrait was perceived as the most unrepresentative of the two groups of highly theoretical and speculative scientists designated as type 4 and type 6. Unfortunately, the scales only constitute a weak validation because (1) of all the scales, the differences among the means of the six different types of scientists are only statistically significant on the first scale, and only just barely significant there, and (2) the order of the six types on each of the four Jungian scales is not exactly in accord with our prior expectations. However the ordering is rather remarkable, particularly when one considers how hard it is to obtain consistent findings in the murky area of personality assessment. Considering this, the agreement among the results of the previous three chapters is rather remarkable.

If the types of this section are representative of some basic attitudes or styles toward the doing of science, and *if* according to

Jung all of the types are potentially possible in any field of endeavor, then the typology of this section allows us to raise some interesting questions for future investigations. Are the portraits developed here sufficient to measure the distribution of types in other fields of endeavor? How would they have to be modified in order to be applicable? Why is the modification necessary? Is the distribution of types the same for all fields? What do the distributions tell us? Are they useful for gaining a novel perspective into the patterns of reasoning and of discovery of various fields of inquiry? For answers we await future inquiries.

Philosophy of Science

As important as it was to make inroads into the conceptualization and measurement of the psychology of scientists, this was not the only area of importance with which this study was concerned. As I have stressed throughout this book, from the very beginning this study was concerned with a host of issues and topic areas. Among the most important of these were the attitudes of the respondents toward some fundamental philosophical and methodological issues of science. We have an enormous literature on what philosophers of science think about the important philosophical and methodological issues of science, and we have the highly select thoughts of outstanding scientists on these issues. But as far as I am aware, we have no systematic studies of what working scientists think about such matters. If such is indeed the situation, then this is truly astounding at this late a date in the course of science. This state of affairs alone constitutes justification for such a study. There certainly is no reason why a beginning in this area could not be made with the particular group of scientists of this study.

Some qualifications are essential before turning to the issues and the attitudes of the respondents toward them: First, it is not the purpose of this book to report on all the philosophical and methodological issues that were posed or even to report any of them in great detail. This would consume far too much time and space. Rather than reporting on all the items, it is more important to report on a few of them to demonstrate the importance of this way of treating philosophical and methodological issues and to give the reader a clear feeling for the responses to a few of the more important items.

Second, it is almost frustratingly impossible to boil complicated

178

philosophical and methodological issues down into the simple attitudinal statements found on most attitudinal instruments. Because of this difficulty I am under no illusion that everyone will agree on (1) my choice of specific items, (2) their wording or boiling down, and (3) the underlying thematic areas that they tap. I am likewise under no illusion that every investigator would have chosen the same issues as I have. Indeed, this raises a very interesting possibility — the greater involvement of philosophers of science in exercises such as this.

My philosophical tradition is that of pragmatism [72, 79, 81, 408], and I believe that one of the important ways to understand the consequences of any philosophical position is to ask how appropriate that position is for all kinds of exercises, not just for the traditional exercises of traditional philosophical analysis. Thus, the philosophical tradition out of which I come argues that the construction of a good questionnaire is just as much a problem for philosophy as it has been a traditional problem for the behavioral sciences. Contrary to the current prevailing mood in philosophy, the construction of a good questionnaire and the study of the beliefs of scientists is not a mere empirical problem or an exclusive problem for psychology or for sociology. Problems are not the exclusive province of any one discipline. Indeed, according to the pragmatist's way of viewing the world, one of the interesting ways to measure and evaluate the contributions of any single discipline is to ask what that discipline has to contribute to the traditional problems of another discipline. This certainly is one way to interpret the meaning and implications of pragmatism's claim that philosophy is a basis for social inquiry, policy design, and social action. That would indeed be an interesting way to evaluate philosophical positions. I for one would be immensely interested to see other philosophical traditions try their hand at constructing attitudinal instruments and exercises on the issues discussed in this section.

Table 14 presents the means and standard deviations of the sample responses to the subset of attitudinal items discussed here. The items are grouped by general topic areas, which are more for organizational purposes than for strict purposes of classification. Many of the items could simultaneously be included under several of the areas. The classification is thus not meant to be exclusive and certainly is not exhaustive.

Although they are informative, the numerical responses (means, standard deviations) to each item as reported in Table 14 are only

179

Table 14

Attitudinal Responses to Some Philosophical and Methodological Issues.

	Mean[a]	Standard deviation
I. Relationship of data to theory		
5. In the long run, science advances more through the doing of critical experiments than through the building of theories.	3.89	1.69
10. Scientific observation supplies us with hard data that are independent of our subjective desires, wishes, and biases.	3.59	2.21
12. Observation of raw data is not only prior to but also independent of theory.	5.00	1.93
32. The dividing line between soft, unproven theories and hard, established facts is actually very fuzzy.	3.81	2.16
33. Theories of science are not tested by data which are arrived at independently of the theory which is being tested.	5.21	2.03
II. Scientific method		
34. A brief account of the *hypothetico-deductive method of science* (H-D) is as follows: (1) Propose a hypothesis as a provisional statement obtained by induction, experiment, or experience. (2) Refine and structure the hypothesis; if possible, state it mathematically. (3) Draw logical conclusions, predictions from the structured hypotheses. (4) Check the predictions against experiment or experience. (5) If the deduced consequences or predictions closely check against the experimental facts, then we are justified in accepting the initial hypothesis as scientifically established.		
(A) The H-D account of SM[b] is, in brief, an excellent account of SM.	3.24	1.74
(B) The H-D account leaves out many essential factors in SM.	3.58	2.01
(C) The H-D account is fantastically naive. What scientists do is vastly more complicated.	4.08	1.92
(D) The H-D account may best describe the finished, end-product of science, but it is a poor description of the process which produced the product.	3.16	1.73

Table 14 (continued)

		Mean[a]	Standard deviation
III.	Metaphysical issues plus general philosophical issues in the testability of scientific theories		
	6. Although science can never prove any theory conclusively true, it can conclusively prove any theory false.	4.21	2.28
	7. Causality is a basic property of the physical world that is inherent in the nature of things themselves.	2.30	1.78
	19. Science does not make metaphysical statements.	3.87	2.25
	20. A single, contrary, "hard" fact that goes against a theory is enough to disprove a general theory.	3.56	2.30
	30. Strictly speaking, causality is *not* a basic property of the physical world that is inherent in the nature of things themselves, but rather, causality is a result of a conceptual decision on the part of a human observer to view Nature as partitioned into "causes" and "effects."	5.00	1.69
	35. The scientist can never subject an isolated hypothesis to experimental test, but only a whole group of hypotheses; when the experiment is in disagreement with his predictions, what he learns is that at least one of the hypotheses constituting the group is unacceptable and ought to be modified; but the experiment does not indicate which one ought to be changed.	4.87	1.98
IV.	Objectivity, rationality, and the social characteristics of scientific knowledge		
	1. Scientific objectivity is guaranteed by the institution and methods of science which subject scientifically generated knowledge to open, public tests by means of rational, impartial test criteria.	2.93	1.72
	2. The logical and rational aspects of science are only a very small part of the total aspects of science.	4.15	1.69
	9. However interesting and rewarding they might be, courses in the liberal arts and humanities really contribute little to the solution of scientific problems.	3.95	2.29

Table 14 (continued)

	Mean	Standard deviation
17. Science is the only field of human knowledge whose results are genuinely cumulative and where genuine progress is made.	5.12	2.16
18. Scientists stick to their pet theories and hypotheses to the bitter end until disgust, frustration and boredom finally make it impossible for them to go on.	4.65	1.63
23. The philosophy of science has little, if anything, of significance to contribute to the actual practice of science.	5.23	1.83
24. Scientists change their scientific opinions more in response to their colleagues' opinions than they do in response to hard, scientific evidence.	4.79	1.58
25. There is much more dogmatism and blind obedience to authority in science than the average person realizes.	3.11	1.81
29. The student of physics really has no need to read the original works of say Newton or Faraday because everything he needs to know about these works is recapitulated in a briefer, more precise, and more systematic form in innumerable up-to-date textbooks.	4.25	2.20
38. The bolder, the more creative, and the more imaginative the scientist, the more likely it is that he will become the captive of his own favorite pet hypotheses and theories.	4.58	1.87

[a]1 = extreme agreement; 7 = extreme disagreement. [b]SM = Scientific Method.

part of the total information associated with each item. First of all, there is a good spread to the distribution of numerical responses to nearly every one of the items, and so the means and the standard deviations alone do not capture the full sense of meaning in the responses. As in the preceding chapter, to get a more accurate overall perspective one must consider the verbal responses in addition to the numerical responses. For another thing, the responses are sensitive to the particular way the issue was posed.

The dominant response to the items under the first topic area is most interesting and significant. Contrary to either a theory-before-data or a data-before-theory view of science [72], it was generally felt that theory and data were so tightly coupled that it was false to think of one before the other or to think of one as more critical than the other; both were critical and so dependent on the other as to be inseparable if not virtually indistinguishable from one another. This sample of working scientists may not have been generally sophisticated with respect to many of the philosophical issues of science. But in my opinion, with respect to the single issue of the relationship of theory to data, they showed a far greater degree of sophistication than the proponents of the orthodox view [129, 180] of scientific theories.

Item #5. Some representative quotes:

Scientist Adams: Science advances equally strongly both ways; therefore I disagree with the statement. I wouldn't rate one more important than the other.

Scientist Baker: I don't know what to mark here because there is no experiment unless there is a theory. The statement is wrong.

Scientist Case: Ah hell! This is the chicken and the egg question. They go hand in hand.

Scientist Davis: You have to have a theory before you can have a critical experiment.

Item #12. Some representative quotes:

Scientist Edwards: I don't believe the observation of raw data is ever really independent of theory. The two things constantly interact. Therefore I'll circle a 6.

The major exception to the dominant response of the interrelatedness between theory and data occurred in the responses to item #33. This item also caused the most confusion of any of the items in the first group.

Item #33. Some representative quotes:

Scientist Ford: I'm not sure what you mean here. My answer is mixed. I believe you can't have data without a model of some sort. Yet on the work-a-day level, you can have a theory and find some data in the archives which settles the question even though the guy who got the data never heard of your theory. I'd agree in the sense that it may not be possible to test the most fundamental theories of science by data which is not somehow dependent on the theories being tested. But I believe this is possible with the more routine, smaller theories.

Scientist Gage: Theories *should* be tested by data arrived at independently of the theory being tested.

The significance and meaning of this item will be more fully discussed in Chapter Seven.

The responses to item #34 show how sensitive and variable the responses are to the wording of the statements. A substantial proportion of the respondents thought that the H-D account was an excellent account, but that it left out many essential factors like intuition. A substantial number also thought that what scientists did was more complicated. But because they reacted strongly to the terms *vastly* and *fantastically naive*, they circled a number towards the disagreement end of the scale. Of all the statements under item #34, (D) produced the most widespread agreement; even those who had earlier tended to agree with (A) and disagree with (B) and (C) tended to agree with (D). As more than one of the respondents put it after getting to (D): "It's been my experience that the scientists who tend to believe this kind of account are the ones who are the last creative. Nobody takes this stuff seriously. You can't put scientific method into a five step procedure." Another scientist put it this way:

> D is probably the best case you can make for this account. The intuition and time element, the feeling that one has to have to get going, are all left out, as well as the technology side of it; so often the experiments one does are those that are easy to do at the present time. At least in my lab a third of the experiments are not hypothesis-free, but they are not definite tests of hypotheses either. They are simply illustrations, hopeful clues, without a clearly formulated process. There is the serendipity business — you're looking for X and you find Y. That has really happened to me and to most people. Technique is very important. All working experimentalists inevitably end up making a very heavy investment in a particular kind of measurement, an investment of a decade or two, trying to be the best in the business. You very definitely tend to do things the wrong way around; problems come up, and you pass them up because you are not good at this technique. You will do a problem inferior to another because it's your bread and butter and you know exactly how to do it. Some people get fascinated by a technique, then, when they get the technique built up, they get fascinated by another technique. That's a very dangerous trip. It's necessary for the working scientist to have an armory of skills (and for the theorist to be good at certain kinds of theory). You're going to end up tackling problems that call for that skill.

Of all the topic areas, the items under III gave the most trouble. Many of the respondents indicated that the terms of the statements weren't clear to them (the largest number of "Isn't clear" responses occurred to these items). Many indicated that they had not thought much about these kinds of issues. As one of the respondents put it, "Look, most of us couldn't care less about these kinds of things; we're too busy doing science to worry about this stuff." The following are some of the more typical responses to the items.

Item #6. Some representative quotes:

> Scientist Hall: I'd agree with the first part of it, but I'd disagree with the second part of it. I doubt you can prove anything conclusively true or false although you can sure marshal one hell of a lot of evidence against something.

The motivation for asking item #6 was to see whether working scientists had any appreciation for the supposed asymmetry between the verification and falsification of scientific theories [36, 353, 461]. In terms of the way the issue is posed in item #6, the answer is no. This is not to say they would not have an appreciation for this distinction if the issue were posed differently or even more precisely.

Item #7. Some representative quotes:

> Scientist Ingram: The term *causality* is not clear to me. This is a very philosophical and abstract idea.

> Scientist Jones: My mind is a blank on this. The notion of cause and effect *is* a basic property of the physical world; it's in the world in that sense. I guess you couldn't have science without a belief in it although I'm not clear what it is.

The respondents seemed to have a basic if not almost primitive belief in the notion of causality and in its operation, although very few were able to articulate this notion in any elaborate sense. The responses generally fell into two distinct classes: (1) the *very* few highly articulate and sophisticated statements on the one hand, and (2) the overwhelming majority of unsophisticated responses on the other.[12] The respondents were generally inarticulate regarding some of the most basic concepts underlying science.

Item #19. Some representative quotes:

> Scientist King: I don't know what metaphysics is.

Scientist Logan: If it is a metaphysical statement, it is not science.

Scientist Meade: Strictly speaking the statement doesn't make sense for logically 'science' can't make statements. Only scientists can make statements. I don't therefore know what you mean. I can't answer the question.

Scientist King's response was the dominant one here, although a number of the respondents used the tactic of Scientist Meade to avoid the question entirely. Extremely rare were more sophisticated responses like the following:

Science does make metaphysical statements even though it doesn't intend to and scientists like to shun them. Science can't help making them, for in the end we are always dealing with things that are never fully understood and hence which depend on fundamental assumptions about the structure of reality. These assumptions are fundamentally metaphysical because you can't prove them; you can't prove them because you would need them in constructing anything pretending to be a proof. So you have to accept or entertain them on grounds other than strict proof or demonstration. You have to accept them as postulates, as grand hypotheses about the nature of the universe.

The responses to item #20 were split along three lines. First, there were those who tended to agree with the statement even though it was with the reservation "if the single fact was 'really hard.'" This first group believed there could be such things as "single, contrary, 'hard' facts." Second, there were those who doubted whether any single fact was enough to disprove any theory, because any single fact by itself could be explained away as an anomaly, that "there are always ways of explaining one fact away." This group tended to question whether there were any such things as isolated, "single," "hard facts." In order for something to be a fact it had to be established and to be established meant that it had to be, if only in principle, replicable; but this meant that it could not possibly be single. It was only a hard fact because it had been shown to stand up through repeated measurement. Finally, there were those who questioned whether any fact, single or not, was ever really hard, because they had more than once seen the hardest of physical facts turn soft under later examination. The responses to item #20 were the most sophisticated of all the responses in this topic area. This may be because item #20 is the most concrete of the statements, the one that most directly relates to the day-to-day working experience of the respondents. It is important to note that, in terms of the notion of various systems of inquiry [72, 306, 314] to be discussed in

Chapter Seven, one would disagree strongly with this item. The notion of a single, contrary, hard fact is nonsensical. Anything purporting to be a fact is the product of enumerable epistemic judgments. These multiple judgments form an integral part of the fact's constitution. In this sense, there is no such thing as a single fact existing by itself. For something to be a fact, one first has to demonstrate the system of inquiry that has produced the purported fact and then subsequently the system of inquiry which is supposed to validate its status as a fact. Further, by themselves, facts — no matter how large their number — are never enough to disprove any theory. It takes a tremendous amount of theory plus facts to disprove any theory [77].

Item #30. Some representative quotes:

> Scientist Nolan: I'm staying away from questions like this one. I don't know what to say about it.

> Scientist Oaks: I have to think about this. I believe in causality, but at the same time I know it is not absolute. Neither pure causality nor pure randomness operates.

> Scientist Park: I think I disagree with this. There *are* causes and effects. I think causality is a basic property.

> Scientist Quay: This is correct but I would go one step further. Not only is causality the result of a conceptual decision on the part of a human observer to view nature as partitioned into causes and effects, but it also is a human conceptual decision to view nature as flowing in time. View here means something like a *Weltanschauung*.

The responses of Scientists Nolan, Oaks and Park were the overwhelming norm here; the response of Scientist Quay was the extremely rare exception. Except for Quay and a very few others, there was literally no appreciation of the Kantian point that causality and, more generally, all of the fundamental notions of science, were not empirical facts or basic structural properties of nature, but instead conceptual creations without which it was not possible to observe nature's facts. For a more extensive discussion of this point, see [72, 79].

Item #35. Some representative quotes:

> Scientist Reed: I don't understand this but if I do, I don't agree with it.

Scientist Smith: I agree with everything up to the final sentence. It's not necessarily always true. The experiment might just indicate that one of the group should be changed. But I agree that a scientist can never subject an isolated hypothesis to experimental test because there are always previous hypotheses whose truth must be assumed. Everything is built on a pyramid so that one scientist's hypothesis is always dependent on another's.

Scientist Tate: No, you can isolate some hypotheses. It depends on the situation. I don't think it's as strong as you've put it.

The responses of Reed and Tate were the dominant ones here; the response of Smith was rarer, but not the extremely rare exception. The philosophically minded will of course recognize this statement as one of the many forms of the Duhem Hypothesis or D-Thesis. In fact, item #35 is a direct quote from Duhem [122, p. 187]. For reasons I have articulated elsewhere [314], I personally agree strongly with Duhem on this point. More to the issue, I believe that one of my colleagues, Laurens Laudan, has made a strong case for the acceptance of item #35 [254].

In many respects, the function of the last group of items was mainly to check on the consistency of the responses of earlier rounds. To a large extent they substantiate those responses. For example, although there was fairly strong and widespread agreement with item #1, there was also strong verbal reaction to it. The general reaction was that it needed strong qualification. A high proportion of the respondents reacted to the word *guaranteed*. They did not know whether there was anything that would guarantee scientific objectivity in actual fact. The most general reaction to #1 was, "This is probably true in the long run, but it is less true in the short run." In this regard, the numerical response to item #25 helps to explain the response to item #1.

The general response to items 18 and 38 was: "Some do; some are like this; but it is too strong to say all are." The implication was quite clear, and in many cases was made perfectly explicit, as to whom the "some" were. Finally, although I do not doubt the sincerity of many of the respondents, I do question the responses of some of them. In terms of their previous responses to earlier items and in terms of their general attitudes on earlier rounds, it seemed that a number of the respondents gave the "socially desirable response" to items 17, 23, and 24.

Concluding Remarks

The concern in this chapter has been with four primary areas: (1) the measurement and assessment of attitudes with respect to some particular scientific hypotheses; (2) the perceived persistence of the commitment of particular scientists with respect to particular scientific issues; (3) the conceptualization and assessment of scientific personality types via an alternate approach; and (4) the assessment of the attitudes of some working scientists with respect to some philosophical and methodological issues of science. With respect to the first area, I have tried to demonstrate that there now exist relatively simple and direct techniques for the measurement of attitudes — and their change over time — with respect to scientific issues. Above all, I have tried to demonstrate the potential benefits of conceptualizing old issues in new ways.

With regard to the first and second areas, I have tried to show that it is possible to make strong statements about the type of scientist that is likely to become committed to particular issues. Unfortunately, from just the knowledge of the specific type that a particular scientist is, it is not generally possible to predict that scientist's position on a specific scientific issue.

In the fourth area, I would say that with respect to some of the basic issues of science, such as the relationship between theory and data, these scientists are rather well informed. However, with respect to some of the most basic philosophical issues of science, they are rather ill informed or unreflective. The distinction seems to be between those issues that are most directly related to their working professional lives and those that are farther removed. But I would venture to say that the same could be said of working philosophers of science. Indeed, the whole point is that complete knowledge regarding science does not reside with one group or the other. Both have serious gaps in their knowledge about science.

Finally, with regard to the third area, I have tried to show that it is possible to make some broad statements about the psychological organization of modern science. Although for the particular group of scientists in this study it was shown that thinking, in the Jungian sense, was the psychological function least present, feeling — in the very special sense with which Jung uses it — is probably the psychological function that has been the least present in the everyday scientific working life of the scientist [100, 102, 106, 215]. The results of this chapter show again a split in the psychological space and life of the scientist. Given the increasing demands being placed on science — for example, that science make explicit its values and value judgments, that it no longer hide behind the myth of a value-free orientation to the world — the biggest challenge may be to show whether it is possible to sweep more of feeling into the working life of the scientist and thereby into the structure of science itself.[13] Feeling has always been concerned with the higher ends and values of all activities. Ethics and values are some of the matters that have always been of concern to feeling [100, 215].

The elucidation and understanding of the psychological orientation of scientists and the reasons for this orientation must certainly rank as an important step in this eventual sweeping of feeling into the structure of science. Given the types that science has traditionally tended to attract and the strong social and psychological forces acting to perpetuate those types, the prospects for accomplishing this are not especially promising. However, from the standpoint of new psychological pressures, especially from within science itself [432], the prospects may be more encouraging. The fact that some of these demands and challenges are being raised by scientists themselves can be interpreted as evidence that new sociological and psychological types, pressures, and movements may be coming to the forefront. However, as the next chapter demonstrates again, the barriers that must be overcome are extremely formidable.

Chapter Six

Science's Moon: Apollo 16

"Representation of the world, like the world itself, is the work of men; they describe it from their own point of view, which they confuse with absolute truth."

Simone de Beauvoir [117, pp. 132–133]

This chapter summarizes the final round of interviews with the scientists. Although a number of issues and areas were explored on this final round, only four are discussed in this chapter: (1) the respondents' judgments of what the significant scientific findings and accomplishments of Apollo were and which scientists, if any, should be given credit for these findings and accomplishments; (2) the assessment and implications of the Apollo missions for the planning of future, large-scale, scientific missions; (3) the respondents' image of the moon in a feeling sense; and (4) the respondents' reaction to this study — what they thought about it and about themselves after having participated in the study.

Significant Findings, Significant Scientists

There were a number of reasons for asking the respondents to list what in their opinion were some of the most significant findings to come out of the Apollo missions and who, if any, were some of the scientists who should be credited with these findings and accomplishments. First, a basic survey of this type is important. It is an indication of what some of the important findings were, and it is also an indication of what a particular group of scientists thought were the important findings. Second, the survey can be used as a way of measuring whether a general consensus was beginning to form in the community regarding the nature of the moon or whether uncertainty, doubt, or conflict was still the rule. As shall be seen shortly, this was an important consideration. Given the strong conflicts of interest and of personality that have been continually investigated throughout this book, it is important to see that there were also scientists who were still able to agree on many things. Third, a list of significant findings is more than just a summary of the scientific results of the missions. It can also be used to infer what criteria the scientists used in judging what was worthy of being called knowledge, of being designated as good science, important science. A list of significant findings and significant scientists is more than just a list of results or a popularity contest; it is also an indicator of group consensus and of the criteria that are used to convert the findings of individuals into the collective knowledge of the community of science.

The findings that were designated as significant break down into essentially four main areas: (1) age dating results, (2) general geochemical and petrological results, (3) seismic and magnetic results, and (4) miscellaneous findings. The single most frequently mentioned important finding was the age dating results, the fact that the ages of the lunar rocks had turned out to be so much greater than what had been anticipated prior to Apollo 11; 33 out of the 43 respondents on round IV, or 77 percent of the sample, referred specifically to the age dating results as significant. An additional measure of their significance was that the age dating results were first on all the lists. However, if one uses the criterion of a general or broad topic area, then the general geochemical and petrological results were the most frequently mentioned findings. Virtually the entire sample in one way or another referred to the geochemical and petrological results. Some merely used the general terms geochemistry or petrology to refer to all the geochemical and petrological findings; others went further and mentioned such specific findings as the general depletion of the volatile elements, the major chemical fractionation of the lunar materials, or the detailed makeup of the lunar soil, rocks.

If we lump them together, the third most frequently mentioned area was seismic and magnetic results; 17 scientists in one way or another referred to the seismic data. The most frequent references were to the unusual or special characteristics of the lunar seismic signals and what this in turn implied about the attenuating characteristics of the lunar materials and ultimately about the structural characteristics of the lunar soil and layers. The fourth and remaining category refers to a relatively small number of infrequently mentioned items, such as the fact that the moon was found to be virtually free of water and the fact that the lunar materials were found to be virtually free of any traces or signs indicative of living organisms.

I will not pursue the list of significant findings in greater depth, because to do so would get us involved in too many technical details and distract us from our main purpose of examining the social implications of the list. The primary questions are: Can we infer a movement toward convergence from the list? Does the list tell us anything about the criteria used in the selection of significant findings and about the scientists who become associated with them? Are there necessarily always particular scientists associated with a piece of significant work?

It is clear from the verbal comments, which were made as the respondents individually listed what they considered as significant findings, that the strong degree of overlap between lists constituted much more than mere chance agreement between abstract lists of physical findings. Although they were still split on many of the technical details and on the interpretation and emphasis to be placed on many of the findings, the marked degree of overlap among individual judgments is indicative of the degree of community consensus that was forming among the respondents about the significant findings of Apollo. From the comments it is clear that many peer and community judgments (standards) entered into the individual judgments. In other words, the individual judgments were not purely individual; they both reflected and helped to create further community opinion. The verbal comments make clear the public nature of the discussion and assessment of the findings that was taking place. In many ways, this was even more apparent from the part of the exercise that asked the respondents if they could associate or credit any particular scientists with the findings.

The verbal comments in response to the assignment of credits make it abundantly clear that there were still intense ideological, personal, and attitudinal differences among the scientists toward some of the issues and toward one another. These differences were manifested, as before, in hostile and critical comments toward one another. In spite of this the respondents were still able to single out those of their peers who were deserving of credit on the issues. And it is more accurate to say "because of this," not "in spite of it." In naming their peers, the respondents were not necessarily or clearly able to put aside their personal feelings. In many cases I saw them using their feelings (as in the preceding chapters) to single out a scientist who was either strongly identified with an issue or finding and who deserved credit for it. It is more accurate to say that the respondents used or overcame their feelings than that they ignored them. Indeed, I also observed a few instances — a *very* few — where some of the respondents could not overcome their feelings. In these few cases it was clear from other comments made earlier that, based on his work, they would have nominated a particular scientist were it not for their personal feelings toward that scientist. Contrary to the popular myth of science that a scientist is supposed to be judged solely on his scientific accomplishments and not on who or what he is, I am not sure this is necessarily dysfunctional. The fact that personal feelings have to be overcome is a check on whether someone is really deserving of credit. As this book has been emphasizing, the involvement of intense emotions is not necessarily detrimental to science. It certainly served to make nomination for credit more than a popularity contest; some of those receiving large numbers of votes were not the most popular.

Twelve individual scientists plus one group, the Preliminary Examination Team (PET) (the first group of scientists to examine the lunar rocks), were singled out for credit for the age dating results. Of the 33 scientists who listed the age dating results as one of the significant Apollo findings, 29 or 88 percent singled out one scientist in particular as deserving of outstanding credit. This was the highest percentage received by any scientist in any area throughout the exercise. Two scientists received 6 votes; one received 4; two others received 2; and 7 others received 1 vote each. It should be noted that there were no restrictions on the number of votes; each respondent could nominate or vote for as many groups and individuals as he wished.

In only two other areas were a relatively large number of scientists singled out for credit. One area was geochemistry; the other was seismic results. Twelve individual scientists plus the PET group were mentioned under geochemistry. Of the 19 scientists listing the general heading geochemistry, 10 or approximately 53 percent voted for one individual; 6 or approximately 32 percent voted for another; 3 other individuals received 2 votes each; and the remaining 7 individuals and 1 group received 1 vote each.

Fifteen scientists listed the seismic results findings as significant. Of those 15, 13 or 87 percent singled out a particular scientist for credit; one scientist received 2 votes; 4 others received 1 vote each.

Although geochemistry was one of the three areas in which the largest number of scientists were singled out for individual credit,

194

it was also the single area in which the respondents commented that it was most difficult or unfair to single out particular individuals for credit. The most typical comment was: "So many of us worked on this and found essentially the same thing that I can't single out anybody." In a sense this same comment applies to all of the areas designated as significant findings, and it thus helps to account for why so few scientists were singled out for individual credit. Because of the large number of scientists involved in the program, it is perhaps rather remarkable that so few scientists were singled out. However, because the sample overrepresents the inner elite of the profession or community, this is not so surprising. Nevertheless, even if one took a larger sample of opinion from those involved in the program, I doubt that the results would change substantially. The reason is that *in every case* those scientists receiving the largest number of votes, no matter what the specific area or category of finding, were precisely those scientists who had been initially identified as the elite of the profession, the type 5s, the outstanding combiners of both theory and experiment, the high status scientists at the high prestige institutions (see Chapters Two and Three).

Although the results are less clear, essentially the same conclusions emerge from three questions about the awarding of prizes for excellent scientific work done in conjunction with the Apollo missions. In addition these questions help to elucidate more clearly the criteria used in the selection of important findings and the assignment of credit to individuals.

The first question was: "If a special prize were to be given for excellent *experimental* work done in conjunction with the Apollo missions, who should get that prize and why?" The responses named 30 separate individuals and groups. The largest number of individuals and groups were named for the age dating results. General geochemistry and seismic results followed. Of the 30 individuals and groups, one individual received 21 votes, the largest number of votes to be cast for any individual. One individual received 5; another 4; six received 2; and finally, 21 received 1 vote each. The individual scientist receiving the largest number of votes here was the same scientist who had earlier received the largest share of the credit for the age dating results. The general accolade indicates the major criteria that were used to single out this individual: "I'd have to single out X for his superior measurements in a technically difficult field. He not only pushed ahead the state of the art before anybody else, but just as important he foresaw the need to do this years before anybody else and planned ahead accordingly. In addition he not only made fine measurements, but he also did a good job of interpreting them." Not all of these criteria were scientific in the strict traditional sense. Although X was strongly credited for his technical accomplishments (improving the state of the art "several orders of magnitude") and for his fine measurements, he was also praised for his accomplishments in putting together a fine scientific team and for his forward-looking planning.

The votes for "excellent theoretical" and for "excellent *over-all* (general)" scientific work were far less clear cut. The verbal comments in response to these two exercises again echo the difficulties voiced earlier with doing outstanding theoretical work in the field of earth and planetary sciences.

Given the perhaps unusual nature of the results of this section, particularly the use of physical findings to infer social consensus and social-scientific criteria, it is important to review and to emphasize some of the major conclusions of this section. (1) If one grants the criterion and the level of consensus that is already implicit in the procedure, then a list of significant findings can be used to infer whether agreement or consensus is taking place within a particular scientific community. (2) In terms of this procedure or criterion, the results indicate that a consensus was forming, at least with respect to the significant findings of Apollo. (3) The procedure along with some associated exercises (for example, the awarding of prizes and the assignment of individual credits) can be used to infer some of the criteria that were being used to select significant findings and to single out individuals. In terms of the last point, the results indicate that the major criteria was that of "outstanding experimental results combined with significant theoretical interpretation." In addition, such organizational variables or criteria as the ability to plan far ahead were also seen as important.

With regard to the importance of empirical considerations as a major criterion in judging the significance of what constituted an important finding, it is worth noting that the vast majority of findings that were listed were of an empirical nature. It is also of interest to report here on the results of an exercise that was given later in the interview. In response to the question about how the respondents would rank the areas in which their understanding of the moon had increased the most, the results in Table 15 were obtained. These results strongly parallel those of the previous

196

Table 15.

Rank Order Indication of Increase in Areas of Understanding As a Result of the Apollo Missions

Rank	Thinking[a]	Intuition[b]	Sensation[c]	Feeling[d]
I	1	12	25	2
II	7	20	9	6
III	18	7	7	11
IV	15	2	0	22

[a] Greater formal understanding of the moon in terms of a formal, mathematical, or physical model,

[b] Greater intuitive grasp or insight regarding the nature of the moon taken as a whole,

[c] Larger store of hard empirical data regarding the moon's detailed and concrete empirical properties,

[d] Greater appreciation for the social and political implications of space exploration and of geology (in the generic sense of the Earth Sciences).

chapter with regard to the distribution of the Jungian psychological functions in the sample. As the scientists said earlier (see Chapter Three), this was the day of the experimentalist.

Finally, although some of the respondents had difficulty in singling out, or were reluctant to single out, any particular individual, on the whole the respondents were easily able to associate particular scientists with particular findings. Even though there were a number of cases where no single individual emerged with a clear plurality of the votes, it still seemed that particular scientists were associated with particular findings. Scientific findings are not impersonal findings; they are associated with their discoverers. Association with a finding, as Merton [294—296] has pointed out, is one of the most significant awards a community can confer on its members.

Evaluation of Missions and Implications for Future Large-Scale Scientific Programs

Table 16 lists the questions that were asked in regard to the evaluation of the missions and their implications for the planning of future large-scale scientific programs.

The responses to item #10 were most interesting. Only item

Table 16

Questions Relating to Evaluation of the Missions

Item Number	Statement or Question	Mean[a]	Standard Deviation[a]
10	How would you rate the performance of NASA in managing and planning the Apollo missions?		
	Outstanding 1 Good 2 Satisfactory 3 Mediocre 4 Unsatisfactory 5 Poor 6 Extremely poor 7		
10a	With respect to management and planning	1.93	1.14
10b	With respect to science	4.50	1.43
10A	What if anything would you have done differently?		
10B	In your opinion, were any serious errors committed in the selection of lunar landing sites?		
10C	Are there any lessons that we should have learned for the planning of future large scale scientific missions?		
11A	Do you think our initial decision to go to the moon is still justified?		
11B	Has it been worth the money?		
11C	What reasons would you now give for justifying our having gone to the moon?		
11D	Are you more convinced than ever of the necessity for manned flights?		
13A	In your opinion, as a result of the Apollo missions, do you feel that the prestige of geology (earth and planetary science in the most generic sense) as a science has increased or decreased?		
	Increased significantly 1 Increased markedly 2 Increased slightly 3 Neither increased nor decreased 4 Decreased slightly 5 Decreased markedly 6 Decreased significantly 7	2.41	1.34

Item Number	Statement or Question	Mean [a]	Standard Deviation [a]
13B	How would you rate the performance of geologists (earth and planetary scientists) as a whole during the Apollo missions? Outstanding 1　Good 2　Satisfactory 3　Mediocre 4　Unsatisfactory 5　Poor 6　Extremely poor 7	2.37	1.20
14	As a result of the Apollo missions, do you have any more or less faith in the ability of geology to be relevant to the solution of man's social problems?		
16A	How do you feel about the decision to include a scientist-astronaut in one of the later Apollo missions?		
16B	Do you think a scientist-astronaut should have flown earlier?		
16C	Do you think it would have made a substantial difference in the scientific progress of the missions if a scientist-astronaut had flown earlier?		
16D	Do you support the choice of the particular scientist-astronaut slated to go?		
17A	How would you describe your performance as a scientist during the Apollo missions? Outstanding 1　Good 2　Satisfactory 3　Mediocre 4　Unsatisfactory 5　Poor 6　Extremely poor 7	2.74	1.11
17B	Did you perform as well as you expected you would? Significantly Better Than Expected 1　Better Than Expected 2　Slightly Better Than Expected 3　As Well As Expected 4　Slightly Worse Than Expected 5　Worse Than Expected 6　Significantly Worse Than Expected 7	3.41	1.81
17C	How well would you say you performed in comparison with your fellow scientists? Significantly Better Than Average 1　Better Than Average 2　Slightly Better Than Average 3　Average 4　Slightly Worse Than Average 5　Worse Than Average 6　Significantly Worse Than Average 7	3.23	1.45

[a] Included where appropriate.

199

#10 as it is shown in Table 16 was presented to the respondents in the questionnaire. A majority of the sample used item #10 to remark that, in terms of management and planning, NASA had done a truly outstanding job and deserved to be rated accordingly. However, if the question had asked about planning for science, they would have rated NASA anywhere from, at best, barely satisfactory or mediocre to unsatisfactory. The feeling about the low priority and emphasis that had been placed on science within the context of the missions was so intense that 10 of the respondents went so far as to write in an extra scale that expressed how they felt NASA had performed with respect to science. Item #10a represents question 10 as it was originally worded. Item #10b represents the write-in question.

The responses to item #10A echoed in good part those to item #10. At least 20 out of the 41 scientists who responded to item #10A indicated, in one way or another, their dissatisfaction with the science planning or component of the missions. The comments ranged as follows: "There should have been more science on the flights." "There should have been more scientist-astronauts flown earlier" (thus anticipating a later question, see items 16A through 16D). "Scientists should have been in greater control." "The program was too engineering oriented." "There should have been more concern with scientific objectives, less playing around on the moon like hitting golf balls, [fewer] conversations with Nixon, and less time spent running the flag up and down." This clearly was the scientists' biggest area of dissatisfaction with the program, so much so that this complaint was constantly echoed throughout the study, not merely on the final round of interviews. Another complaint that was also heard frequently during the interviews was that there should have been more time between missions, that the missions were too closely spaced for maximum scientific effectiveness in analyzing the returned rocks and for further planning (see Chapter Three for a previous discussion of this point). A number of the respondents felt that the sites were too rigidly fixed in advance, that the selection of the sites was not flexible enough to take into account results from previous sites. A number of the respondents also mentioned that the Apollo missions should have at least been continued up through 19 or 20 and possibly extended up to around 24. As it was put by more than one respondent, "It seems foolish to have invested all that money for a Rolls Royce [the start up costs of the Apollo program] and then not have the gas to run it."

The most reflective comments in response to item #10A were the following:

> More diverse scientific schools of thought should have been involved in the planning of the missions. Too much of one point of view dominated. It was also a mistake to reduce the contribution of geology to the level of an experiment. Geology has more to offer than tightly controlled experimentation. The selection of PIs, of tight experiments, forced the work to be too specialized; it inhibited the development of a comprehensive picture of the returned samples. Such a picture does not yet exist.

The responses to item #10B were about equally split three ways among (1) those who felt the sites were all right, that in hindsight we had chosen pretty well, (2) those who did not know and felt that in hindsight one could always criticize anything, and (3) those who felt the site selection was poor, that there should have been more highland sites. The most widespread feeling was that there were too few missions.

Most of the responses to item #10C were essentially the same as those to 10A and 10B: More diverse disciplines should have been involved in the planning of the missions; there should have been more missions; there should have been more flexibility in the scheduling of the missions; there should have been more emphasis on science, and so on. Some of the new responses included: "There should have been a greater mix of manned and unmanned missions; it was a mistake to put all our emphasis on manned missions; once the Apollo missions are over we have no unmanned capability like the Russians for continuing lunar exploration."

A few of the respondents also mentioned that we should be aware in future missions of the need for "selling science to the public and politicians," "of the necessity for public relations on an international basis for projects of this kind." I would say that of all the responses to item #10C, the most reflective responses were those that argued that "a greater mix of scientific disciplines were needed in planning the missions," "that we shouldn't decide too soon as to which disciplines should be dominant." Other than this, I do not consider the responses particularly novel or enlightening. Considering the cost of the program, one would have hoped for more reflective responses. We might have learned more regarding the nature of science planning, and especially from this particular group of scientists as one of the most important of the groups involved in the planning of the Apollo missions.

Questions 11A through 11D are clearly four of the most important questions we could ask of any group that was as deeply involved in the Apollo program as these scientists were. The decision to go to the moon has been subject to incisive analyses and critical attacks [44, 484, 485], particularly from the standpoint of the tremendous social costs and deferred social opportunities. And so, if only for reasons of comparison, it is important to find out how a group of influential scientists feel about these matters.

Of the 41 scientists who responded to question 11A, 38 gave simple, direct yeses. Even more striking was the fact that 32 out of the 38 who responded with a yes gave no further qualification or explanation of their answer. This was quite surprising even given the fact that question 11C, which was clearly visible a few lines below on the questionnaire, called for an explanation. My distinct personal impression was that unless question 11C had been deliberately included, I might not have gotten any further response. Because there were so few verbal responses other than yes or no to question 11A, I can list almost all of them.

> Yes, but we shouldn't have done it in such a crash way.

> Yes, in terms of scientific rewards.

> Yes, from both the scientific and the standpoint of national pride.

> Yes, in terms of the knowledge gained.

> Yes, but only on political grounds; an unmanned program at one-tenth the cost could have been justified.

> It never was.

> Scientifically, no; robots could do it for one-twentieth to one-fiftieth the cost; for imagination, it was fun.

The responses to question 11B were as abbreviated. Of the 40 scientists who responded to 11B, 35 gave yeses, and only 8 of the 35 expanded on their response.

Although not as brief or as few as the responses to question 11A and 11B, the responses to question 11C were nevertheless not so varied or so extensive that they cannot be easily categorized by theme. The responses essentially break down into three main categories: (1) curiosity and adventure, (2) politics, and (3) scientific knowledge. With respect to the first category, the responses were: "to satisfy our curiosity," "exciting," "it was a supreme challenge to our science and technology," "because it's there," and "man's future in space."

202

With regard to the second category, the responses were: "for reasons of national prestige," "as a stimulus to the economy," "national pride," "as a back-up for national defense," "competition with the Russians," "the achievement of parity with the USSR in manned space flight using the landing of men on the moon as a spectacular national goal," "work for the aerospace industry," and "its value as a great project around which to unite the people of the country."

And finally, with regard to the third category, the responses were: "to increase our scientific knowledge about the moon and earth, the solar system," "new knowledge of the early evolution of the earth from a study of the moon," "to place all of planetary science on a more advanced plane," "because there are first rate cosmological questions involved," "to develop the technology of operating in space."

In terms of the extensive criticisms that have been made of the program, I would have to judge the responses to questions 11A through 11C as woefully inadequate and extremely disappointing. They barely even begin to address themselves to the range of serious issues involved; those few that do are at such a superficial level as to be cliches. I think it is terribly important to mention that throughout the study a good many of the respondents continually rationalized their position with the following line of argument:

> Ideally one could argue that this money should have gone into social programs, the rebuilding of the cities, and so on. However, there is no reason to assume that if we hadn't spent the money for Apollo it would have gone into these programs just [as] there is no reason to believe that if we hadn't had Vietnam we would have directed that money into doing social good. Things don't work that way. Besides, the problems of the cities are so expensive that a few billion dollars wouldn't even begin to make a dent in them. At least we accomplished something significant and lasting with this money. The nation got its dollars' worth with the money invested in Apollo.

I cannot really add anything in the way of criticism of this line of argument that I haven't already said in Chapter One — I think this is always the technologist's rationalization and self-fulfilling prophecy. He always argues that social programs are costly, and that personally *he* supports them, but that they require widespread public support if they are to be effective. However, because such programs rarely have widespread support, and because the amount of money is always limited, he can always argue that it is more effective to spend the money in his area where he can objectively demonstrate accomplishment. The flaw in the argument, of course, is that the technologist is one of the major contributing factors to the lack of widespread support. His very position or policy diverts needed public energy and attention from the messy social problems back here on earth. What is particularly infuriating is that the technologist does not mind relying on and actively molding widespread public support, especially when it is to his advantage. When it is not, he argues that it cannot be molded, that no matter how much money you put into a particular problem, it will never be enough to make headway on the problem, let alone to solve it — as though one should only put money into social problems when one is assured of solving them.

The only difference between this enactment of the technologist's rationalization and previous plays is that this time it was enacted on the cosmic scale — it was truly geopolitics writ high in the sky. It is not that I believe we should not have gone to the moon. But, as I discuss in the next section, the point may be *how* we should have gone, with what psychological spirit and frame of mind we should have gone.

With regard to the issue of manned versus unmanned space flights (item #11D), 24 of the respondents were still in favor of manned flights; 13 were not; and 4 were neutral or had no opinion. A number of the scientists — particularly those who had initially supported a strong manned program — now felt that the best thing would be a good mix of manned and unmanned exploration. However, a clear majority of the sample felt that a manned program was essential in the beginning and had clearly demonstrated its superiority to the Russian unmanned efforts; at this point in their development, machines just could not collect complex samples as men walking on the surface of the moon had done.

Items 13A and 13B require little elaboration. The majority of the respondents felt that the status of their science as a science had clearly increased. Likewise they felt that, on the whole, their performance as scientists had been good.

The motivation behind asking question 14 was to get a check on some of the earlier responses to questions like 11A, 11B, and 11C. And the form of the question was used to deliver a strong challenge to those responses. One of the many arguments that NASA has always used to justify support of the moon program is that the techniques and technology involved in getting to the moon might be useful in providing novel solutions to some of our major social

204

problems — such as pollution, crime, and transportation — back here on earth. Question 14 is not basically concerned with whether the moon program actually has spun-off useful solutions to our social problems. Rather, it is concerned with whether, as a result of the program, the respondents have a changed conception of the relevancy of their science to social problems. Of the 41 responses to question 14, 8 scientists answered that geology was more relevant to the solution of man's social problems; 10 answered that it was neither more nor less relevant; 8 answered that geology was irrelevant; 3 answered that it had no relevancy at all; 1 answered that it had less relevancy than before; 11 made no comment. Those who answered that geology was more relevant mentioned such things as geology's importance in uncovering and developing our natural resources. Two scientists mentioned that the greatest benefit of the space program may have been in persuading people to take ecological problems seriously, that the view of the earth from the moon may have gotten across to people for the first time the beauty and fragileness of our planet. For the first time people could see that we all lived together on one single body. In this and other senses, one of the respondents never doubted that geology was always relevant to social problems. Unfortunately the majority of this scientist's colleagues did not see it in the same way. Some saw the Apollo missions as either unrelated to social problems or of "no correlation." Some said, "Social problems are a separate problem." Most did not know. The particular scientist who saw geology as less relevant replied, "Geologists have a piece-meal approach to complex problems which precludes their solving them."

In response to question 16A, 33 of the respondents felt that a scientist astronaut should have flown earlier. Some of the comments were: "There should have been a scientist astronaut from the start." "It was essential; it's almost too late now." However, in response to the direct question of item #16B, only 25 (but still a substantial number) of the respondents felt a scientist astronaut should have flown earlier on the missions; 6 said no flatly; 3 said maybe; 2 said not necessarily; and the rest had no strong feelings. Even among the 25 who said yes, there were still some strong reservations such as, "Only if he was a good astronaut," and "Some of the astronauts have done some good science." There were also some strong feelings in favor of scientists, such as, "It's easier to train a scientist to be a test pilot than to train a test pilot to be a scientist."

Of all the responses to the various questions under item #16, the responses to 16C were the most fascinating and important. The sample was nearly equally divided on this question. On the one hand were those who wanted a scientist-astronaut on board precisely because they wanted a trained scientific observer. On the other hand were those who did not want a scientist-astronaut on board precisely because they did not want a trained scientific observer. Those who favored a scientist-astronaut were disturbed by the "lack of perceptive observations by the astronauts," because "the observations of the astronauts were uncontrolled and unsystematic." Those who were against a scientist-astronaut were afraid that a scientist-astronaut's observations would be too controlled, too systematic, biased and colored by his scientific point of view. As one of the respondents for this position put it:

> Look, it's a matter of choice; with a scientist you get his preconceptions of what's important to collect. Everything he does is colored by his geologic experiences of Earth. But it's a whole new ball game on the moon. Maybe the one thing you don't want is an experienced observer. I think I'd prefer to go in this case with the relatively unsophisticated observer with relatively few built-in biases or preconceptions. I don't want mine or anyone else's biases collecting rocks on the moon. I heard that Smith wanted one of the astronauts to pick up some Y rocks to prove his point. I got madder than hell when I heard that.

This difference in opinion is fascinating, particularly when one considers that one of the fundamental questions in science is about the role of or the necessity for the trained observer. Science considers the trained observer an absolute necessity, so much so that it has evolved long and careful procedures for his training in every branch of science. Yet we find here a substantial proportion of a group of scientists arguing for the benefits of the untrained observer. In either case, the effect of observer bias is attested to. It is impossible to observe without preconception [168, 170]. The trade-off is not between preconceptions or none at all, but between one set versus another. Evidently a substantial proportion of the sample preferred the preconceptions of an astronaut to those of one of their own. The comparison of this response with those in Chapter Three arguing the virtues of commitment and bias is most interesting.

With reference to item #16D, the vast majority of respondents were strongly in favor of the particular scientist astronaut slated to go. He received much strong praise. However, many of the respondents felt that it would be hard to beat the Apollo 15 astronauts, that their performance was so good as to virtually equal that of a scientist.

Moon of Man

In *Of a Fire on the Moon* [270, p. 410], Norman Mailer wrote, "If the moon was not sinister, then NASA was heir to a chilling disease, for they had succeeded in making the moon dull." In *Woman's Mysteries Ancient and Modern* [172, p. 20], M. Esther Harding wrote: "The symbol which above all others has stood throughout the ages for woman, not in her likeness to man, one aspect of *homo sapiens*, but in her difference from man, distinctively feminine in contrast to his masculinity, is the Moon. In poetry, both modern and classical, and, from time immemorial in myth and legend the moon has represented the woman's deity, the feminine principle. To primitive man and to the poet and dreamer of today the sun is masculine and the moon feminine."

When social scientists, historians, and philosophers study science and scientists, they very rarely study what scientists feel about the objects of their investigations. Most typically they study what scientists think about the nature of the moon, life, the universe, and even about their fellow man considered as an object for scientific scrutiny. As I indicated in the previous chapter, science and even the study of science have become synonymous with thinking. Up to this point we have examined what the scientists of this study have thought about the moon from a scientific point of view. There has also been an examination of what they felt and thought about science and their fellow scientists. What has not been done is to examine how they felt about the moon. The moon is more than just an object of scientific inquiry. Long before it ever was an object for natural inquiry, it was an object for some of man's deepest feelings, a cultural object, a symbol onto which man projected some of his most ardent desires, dreams, fantasies — an object of myth. The question, then, is what meaning, if any, the moon has for our scientists in a mythical sense.

A semantic differential on the moon as a cultural object was formed and administered in accordance with the discussion and procedure of Chapter Three. Both quantitative and qualitative (verbal) responses to the various scales were obtained. That is, the scientists were encouraged to verbalize freely in response to each of the scales. In this way the scales became projective devices in

Table 17.

A Semantic Differential on The Moon As a Cultural Object

Scales		The Moon	Scale Means	Standard Deviations
(1)	complex	simple	2.93	1.66
(2)	hospitable	inhospitable	5.00	1.96
(3)	masculine	feminine	3.84	0.85
(4)	depressing	exhilarating	5.93	1.10
(5)	awe-inspiring	commonplace	2.24	1.53
(6)	conquered	unconquerable	3.64	1.37
(7)	astronomical	astrological	1.51	1.21
(8)	animate	inanimate	6.00	1.81
(9)	naturalistic	mythological	1.74	1.08
(10)	disclosing	secretive	3.15	1.51
(11)	poetic	scientific	5.54	1.14
(12)	easy-to-understand	difficult-to-understand	4.76	1.56
(13)	interesting	dull	1.40	0.70
(14)	concrete	elusive	2.42	1.31
(15)	romantic	realistic	5.15	1.48
(16)	dead	alive	3.79	2.03
(17)	scientific	technical	1.97	1.20
(18)	as separate parts	as a whole body	4.77	1.78

addition to being quantitative devices. Table 17 gives the quantitative scale responses and the semantic profile formed by connecting the means of the scale values. For the most part both the scales and the responses are self-explanatory and require little comment. Instead of discussing all of them I shall concentrate instead on one scale. More than any other, scale #3 captures the dominant underlying response to the whole exercise. The dimension is masculine —feminine.

Of the 42 scientists who responded to scale #3, 23 marked the value 4 and 10 gave no response at all — the largest number of "no responses" to any of the scales. More important than the numerical responses alone is what they signified. The verbal responses clearly and overwhelmingly indicate that 33 of the scientists (whether they marked 4 or left the scale blank), or 78 percent of the respondents, had no feeling for this particular dimension; they were unable to take it, to respond to it, very much in the manner that McClelland [285] notes that a substantial proportion of creative physical scientists are unable to take the TAT (Thematic Apperception Test). The overwhelming verbal responses to this scale were: "This scale is irrelevant. I don't think of the moon in these terms. I have no feeling for this dimension. The moon is neither masculine nor feminine to me."

These comments were delivered with great passion by many of the scientists. A number of the respondents were quite intense in their feeling that the moon was neither masculine nor feminine. *They were not neutral about their seemingly neutral response.* Another indication of this strong feeling was that this particular scale elicited the largest number of guffaws, smiles, and humorous reactions. By now there should be no need for us to dwell on the potential psychological meaning of these responses. I have already pointed out repeatedly the intense masculinity of this group and of other groups of scientists, and we do not have to retrace the arguments leading to this interpretation of the responses [285]. This facet of the scientists' psychology emerges again and again, no matter what the exercise.

In *Of a Fire on the Moon* [270], Norman Mailer has argued that it was the Wasp who in spirit and in body took us to the moon, who literally went to the moon: "The real function of the Wasp had not been to create Protestantism, capitalism, the corporation, or a bastion against communism, but that the Wasp had emerged from human history in order to take us to the stars. How else to account for the strong, severe, Christian, missionary, hell-raising, hypocritical, ideologically simple, patriotic, stingy, greedy, technology-deploying, brave human machine of a Wasp? It was a thought with which to begin to look at astronauts [270, p. 316]."

If Mailer can argue this, and I not only think he can but that he should, then I think I can argue with Esther Harding, because of the semantic differential data, that it was man, not mankind, who in body, spirit, and soul took us to the moon, who landed on the moon, who took back some of that precious moon, and finally who analyzed that moonstuff. Nowhere in all of this was the feminine principle present. It was no accident that the first ship that landed was named Eagle. The landing of men on the moon constituted the supreme insult to women, not because men alone went to the moon but because they went with so little sensitivity, with so little appreciation and respect for the moon as the supreme symbol of the feminine principle. We still do not recognize this. Men went with so little understanding and appreciation that the poet's vision of the moon (see, for example, W.H. Auden's [450, pp. 21—22] poem "Moon Landing") is deservedly different from that of the scientist, the male. This vision of the moon should have gone along as an integral part of the planning and execution of the missions. We still do not realize that masculinity is not and should not be confined solely to men, nor femininity solely to women, but that there is a feminine and a masculine side to each of us [208, 213—216]. Our science is predominantly masculine, with little appreciation for the feminine [70]. We must recognize the challenge to learn how to do science with feeling, how to develop a science that in its working methodology and spirit knows what feeling means [309]. This is what we should have learned for "the planning of future large-scale scientific missions," — how to achieve a true mix of disciplines and moods in the planning of the missions. I am not saying that men should not have gone to the moon, or that our science should not be masculine. Rather, I am saying that men alone should not have gone and that our science should not be exclusively masculine. It should be feminine as well. Certainly, one way of looking at the world is not superior to the other. Our science and our poetry, our concepts of masculinity and of femininity, are all deficient and incomplete because they are so one-sided. As Esther Harding has put it: "We have given our allegiance too exclusively to masculine forces" [172, p. 241].

As a final brief commentary on the degree to which the spirit of the program was under masculine domination, it is worth examining the symbolic makeup of the mythic hero Apollo chosen as the official name of the program. Although there may be no standard interpretations of the symbolism of heroes as there presumably are in science, I defer to Norman O. Brown in these matters. If Brown is to be taken seriously, then the implication to be derived from his argument is that man went to the moon not only in body and spirit but in name, both actual and symbolic, as well. The very name that was chosen to symbolically represent the spirit of the flight was masculine. I quote from Brown:

> To understand our present predicament we have to go back to its origins, to the beginning of Western civilization and to the Greeks, who taught us and still teach us how to sublimate, and who worshiped the god of sublimation, Apollo. Apollo is the god of form — of plastic form in art, of rational form in thought, of civilized form in life. But the Apollonian form is the form as the negation of instinct. "Nothing too much," says the Delphic wisdom, "observe the limit, fear authority, bow before the divine." Hence Apollonian form is form negating matter, immortal form; that is to say, by the irony that overtakes all flight from death, deathly form. Thus Plato, as well as his shamanistic predecessors Abaris and Aristeas, is a son of Apollo. Apollo is masculine; but as Bachofen saw, his masculinity is the symbolical (or negative) masculinity of spirituality. Hence he is also the god who sustains "displacement from below upward," who gave man a head sublime and told him to look at the stars [53, p. 174].

Brown does not stop here. He is not merely content to point out the limitations of Apollo, but as is well known, he is also concerned with suggesting an alternative as "a way out of a morass of sublimation that we know as civilization." Brown leads us to the suggestion that if we had been truly concerned with going to the stars in the name of all mankind, and not merely in the name of half of it, we could have chosen a better hero than Apollo: "But the Greeks, who gave us Apollo, also gave us the alternative, Nietzche's Dionysus. Dionysus is not dream but drunkenness; not life kept at a distance and seen through a veil but life complete and immediate..... Instead of negating, he affirms the dialectical unity of the great intellectual opposites: Dionysus reunites male and female, Self and Other, life and death" [53, p. 175].

It should be clear by now that I am not quibbling about something as trivial as what a thing shall be called. Rather I am concerned with the symbolic content of science. As Albert Szent-Gyorgyi put it in a recent letter to *Science*:

> In science the Apollonian tends to develop established lines to perfection, while the Dionysian rather relies on intuition and is more likely

to open new, unexpected alleys for research. Nobody knows what "intuition" really is. My guess is that it is a sort of subconscious reasoning, only the end result of which becomes conscious.

These are not merely academic problems. They have most important corollaries and consequences. The future of mankind depends on the progress of science, and the progress of science depends on the support it can find. Support mostly takes the form of grants, and the present methods of distributing grants unduly favor the Apollonian. Applying for a grant begins with writing a project. The Apollonian clearly sees the future lines of his research and has no difficulty writing a clear project. Not so the Dionysian, who knows only the direction in which he wants to go out into the unknown; he has no idea what he is going to find there and how he is going to find it. Defining the unknown or writing down the subconscious is a contradiction in absurdum. In his work, the Dionysian relies, to a great extent, on accidental observation. His observations are not completely "accidental," because they involve not merely seeing things but also grasping their possible meaning. A great deal of conscious or subconscious thinking must precede a Dionysian's observations. There is an old saying that a discovery is an accident finding a prepared mind. The Dionysian is often not only unable to tell what he is going to find, he may even be at a loss to tell how he made his discovery [429, p. 966].

Taking Leave

If social scientists rarely ask their subjects about how they feel about the objects their subjects study, then social scientists just as rarely ask their subjects about how they feel about participating in social science studies, about how they feel about being an object of investigation from the social scientist's point of view. Below is a list of the final set of questions that were asked on the final round of interviews. Growing numbers of researchers are beginning to appreciate the importance of asking such questions [194, 405]. They are important for getting a check, no matter how rough, on the reliability and validity of previous answers. Even more important, they at least make the attempt to get at the frame of mind with which the subject approached the experience, and to ask what he got out of it.

Now that this study is at its end, I'd like to ask for some of your reactions to it. (1) Have you enjoyed participating in this study? Have you found it interesting? Why or why not? (2) Have you talked to others about this study? If so, may I ask, who and what you said to them? (3) From your point of view, what is the primary thing you think I've been studying? (4) Has this study made you think about anything concerning science that you haven't thought about before? (5) Has the fact that you've been interviewed made you more conscious about any of the things you might have taken for granted before? (6) Has the study made you more consciously aware of the behavior of your fellow scientists? (7) Has the opportunity to talk about the nature of your work and field to someone who was not directly involved with it been of benefit to you in any way? (8) Have you learned anything of benefit to you by participating in this study?

With regard to question 1, even though I purposefully asked the respondents to be brutally frank and blunt in their responses, I have no sure way of knowing whether they were. In fact, I suspect a number of them were overly kind. Nevertheless, 36 out of 43 answered yes; 2 answered no, and the rest either gave no comment or adopted a wait-and-see attitude. More important than the positive and negative answers were the reasons.

It forced me to state my opinion on controversial topics.

It needed to be done and no one with the proper background has done this kind of study before.

I enjoy analyzing the motivations and judgments of my fellow scientists.

I thought the questions were thought provoking.

The whole thing was just fascinating.

People are as interesting as the moon.

It helped me to think about myself.

Because the social sciences are more important and study more difficult things than the physical sciences.

Because scientists are important in our society; therefore it is necessary to understand why and how they do the things they do.

I am intrigued by the thought processes of scientists and would like to understand them better.

Because it's a well constructed effort to get at the subjectivity inherent in the judgments of scientists.

Because we all like to gossip.

Those who responded no and those who gave no comment were unanimous in their feeling of wait-and-see; many of those who responded yes also shared this feeling. They wanted to reserve judgment of the study until the finished product appeared. Until then, they would not commit themselves. It was not enough just to evaluate what they had seen thus far. It was extremely difficult for me to keep coming back to the same group and to hold off giving in to their repeated requests for results. There were a few respondents in particular whom I literally had to fend off with every trick and device at my disposal to keep from giving them any preliminary results. From other exercises I have reason to believe that this behavior was not expressed only toward me, but rather that physical scientists are extremely achievement or product oriented.

Nineteen scientists said they had talked to some of their colleagues about the study and the fact that they were participating in it. However, from the comments it was clear that on the whole they had not talked much about it among themselves, and that when they did the discussions were very brief and general. Even more interesting was the differential perception of who talked to whom and what constituted "talking to someone else about the study." For example, Scientist X would mention that he had talked to Scientist Y about the study, but when Y was queried, Y didn't mention or recall that he had talked to X at all. The remaining twenty-four respondents indicated that they had not talked to anyone else about it.

Here are some typical responses to question 3:

The nature of scientific discovery.

The evolution of scientific consensus.

The evolution of scientific ideas.

The personality of scientists and its relationship to their work.

Hopefully the way scientific hypotheses change in the minds of scientists as data develop.

How people act as scientists and how science has gotten as far as it has with humans doing the work.

Scientists and how they really work.

How scientists function in the typical situation involving unusual publicity [sic].

Me.

214

Fourteen scientists answered yes to question 4. Here are some of the responses:

The more creative aspects of science.

The interactions between scientists and the role of scientists in society.

It's made me more aware of the pervasiveness of extreme thinking and behavior on fuzzy issues like the origin of the moon.

It's made me more aware of how ideas change, fadism in science; it's also given me a greater respect for the social sciences.

It's really made me articulate old ideas more than think about new ones.

Twenty-nine scientists answered no to question 4. The dominant response of this group was:

I always had these insights; I always knew science wasn't as the textbooks described it.

Sixteen scientists answered yes to question 5. Some of the responses here were:

It made me more aware of commitment as a force in science.

It forced me to consider the qualitative aspects of my work.

It made me particularly aware of the differences between certain scientists by being forced to draw comparisons between them.

I was always aware of these things; the study did make me more interested in them.

The remainder of the sample answered no to question 5. The dominant response here again was that the study had not made them consider anything that they were not already aware of.

The responses to question 6 were substantially the same as those to question 5. Most of the respondents indicated that they were already too aware of the inflexibility of some of their peers.

Twenty-seven scientists answered yes to question 7. Some of the responses here were:

It was like going to a shrink.

It was of psychological therapeutic value [two scientists gave this response].

It made me look at my ideas somewhat more critically.

Sure, and often more than talking to a fellow scientist.

It caused me to reexamine my motives in pursuing scientific research.

At least it didn't harm me.

I enjoyed you personally and as well as the opportunity to spread some of the geologic gospel.

The remainder answered no in one way or another, e.g., "not scientifically," "not particularly."

Fifteen scientists answered yes to question 8:

More awareness of the people involved in scientific endeavors.

Better appreciation for the scientific status of sociology and group psychology.

To be more wary.

It made me more aware of the difficulties of devising an ideal survey of this sort.

Better understanding of the social context of science.

Two scientists answered don't know, 6 answered not yet, 2 answered not much, 2 answered that they hoped they would soon, 8 flatly answered no, and there were 8 no comments.

216

PART III

Theoretical Endings

Chapter Seven

Objectivity in Science

"A real rejuvenation of science would have to be based on a radical modification; it would be based on a new ideology of science, and realized in debate and in institutional struggle."

Jerome Ravetz [356, p. 36]

"Tomorrow's science will not be objective. Rather, the dichotomies, objective-subjective, biased-unbiased, will cease to have their present significance. Future science will not be politically immune."

C. West Churchman [77, p. 209]

One of the key questions that the results of the preceding chapters clearly raise is the status of the important notion of scientific objectivity. If scientific knowledge is the product of committed observers, and perhaps more than we realize, the result of observers with strong biases, then the fundamental question is whether objective knowledge is possible in science at all.[1] At their face value most of the major results and conclusions of this book would seem to spell a death blow for scientific objectivity as it has been traditionally conceived, if not for the very notion of science itself. And in a sense this is correct. But in another sense it depends on which notion of scientific objectivity one is talking about. The results of this book are clearly not compatible with every notion of objectivity. However, the more interesting question is what notion of objectivity, if any, the results of this book suggest. In other words, how is scientific objectivity possible, not in spite of but taking account of and building on the results of this book? The fundamental question is how scientific knowledge is possible.

Any theory of scientific objectivity — of science in general — must account for at least the following: 1) there is conflict between two sets of opposing norms of science, and scientists recognize the operation and validity of both sets of norms in their attitudinal responses; 2) there are distinct styles of inquiry (Inquir-

219

ing Systems, [72, 306, 314]) in science, as well as distinct psychological types of scientists, and scientists vary markedly in their appreciation and tolerance for these different styles; and 3) although scientists critically test their ideas, they do so through an adversary proceeding that basically combines deep formal elements (i.e., philosophical systems and methodological standards) with intense informal elements (i.e., psychological and sociological processes). Any theory of science which ignores these crucial factors is doomed to incompleteness and ultimate failure of explanation.

At this time I seriously doubt our ability to construct a comprehensive theory of science that is able to do equal justice to all of these factors. Neither our philosophical, psychological, nor sociological theories are well developed enough to show the simultaneous and detailed operational impact of all these factors on one another, let alone the combined impact of all three disciplines on one another. The purpose of this chapter is therefore the more modest one of suggesting a framework — a program of research — that will hopefully lead to the establishment of a combined philosophical, psychological, and sociological theory of science. The goal is the suggestion of a framework wherein a theory of scientific objectivity, which explicitly takes account of the results of the preceding chapters, might still be possible. The purpose of the final and concluding chapter is to show how the results of this book bear on two specific philosophical views of science that are of current interest, those of Thomas Kuhn and Paul Feyerabend [132—135, 246—248].

Epistemology as Systems of Inquiry

Concepts can never be defined in isolation from one another. Whether explicitly or implicitly, the definition of any term always presupposes the definition of a system of other terms [16, 72]. Granting this, it is helpful to approach consideration of the notion of objectivity through a perspective that makes explicit the use of various kinds of systems. For one, it will be shown that there are at least as many conceptions of objectivity as there are fundamental kinds of systems known as Inquiring Systems (IS) [72, 306, 314].

The systems that will be examined here come from the history of Western epistemology. Epistemology is used because: (1) it represents the most general attitude of man toward the problem of how to build models (conceptualizations) of the system called the *real world*; (2) the history of epistemology is anything but unanimous in its choice of a single, best model; and, hence, (3) a survey of Western epistemology provides us with the broadest possible survey of the most disparate, widely conflicting attitudes that man has taken toward the problem of conceptualizing fundamental models.

The models considered here derive from the recent efforts of C. West Churchman to formulate some of the major systems of epistemology in such a way that they could become of direct relevance to the information-systems needs of the practicing scientist. The title of Churchman's effort, *The Design of Inquiring Systems* [72], is meant to emphasize that to conceptualize or to model a problem *is* to conduct an inquiry into its nature, and that to conduct an inquiry is to gather or produce some information on the nature of the problem. In this sense, *information* is a direct function of *epistemology*. What we know about a problem (i.e., the information we have on its nature) is a function of how we have obtained that knowledge, i.e., of some system of inquiry. Because information is such a strong function of inquiry, *The Design of Inquiring Systems* is an exploration into the design of archetypal philosophically-based information systems. To model a problem is to present information on its nature to some decision maker who is (or may be) required to take action on the problem [16, 70, 81]. Of necessity the discussion here of each inquirer must be brief. For more extensive discussions the reader must be referred to Churchman [72, 73, 79] and to papers by this author [306, 314].

Neither Churchman nor I is claiming that these systems are exhaustive of the class of philosophic systems. Such a claim would be as absurd as it would be pretentious. We are merely claiming that these inquirers are representative of some basic attitudes toward modeling. Further, we are not claiming that everyone would agree with our labels for each system. The labels represent Churchman's characterization of the major spirit of each Inquiring System (IS) and of the historic system to which each most nearly corresponds. This is an exercise in systems design and analysis, not in philology or etymology.

Given a problem, each IS will, in general, produce a radically distinct representation or conceptual model of it. The reason is that each inquirer starts from radically distinct types of fundamental building blocks, i.e., primitive elements or "elementary units of information." In addition, each inquirer embodies a radically distinct kind of *guarantor* for insuring the validity of the "final information content" that is built up from the elementary building blocks. As a result, the final information which is generated from the blocks will be characteristically different for each IS. In effect, the final information content of an inquirer with respect to a problem is that inquirer's representation of that problem.

The Inquiry Systems (IS) that will be examined here are: (1) Leibnizian or Formal-Deductive IS; (2) Lockean or Consensual-Inductive IS; (3) Kantian or Synthetic-Representational IS; (4) Hegelian or Dialectical (Conflictual) IS; and (5) Churchmanian-Singerian or Pragmatic-Interdisciplinary IS. Those readers who find such terms as Leibnizian and Lockean at variance with the meanings to which they are accustomed may find it helpful to think of these systems in terms of the labels Formal-Deductive, Consensual-Inductive, and so on.

Leibnizian IS

Leibnizian IS are the archetype of formal-deductive systems. Leibnizian IS emphasize the purely formal, the mathematical, the logical, and the rational aspects of human thought. They represent the side of scientific inquiry that has always been interested in the construction and exploration of purely theoretical models. They also represent the attitude of scientific thought that perceives the construction of comprehensive, abstract theories as the supreme achievement and ultimate goal of all science. The essential characteristics of Leibnizian IS can be captured as follows: Leibnizian IS start from a set of (1) *elementary, primitive (i.e., undefined) explanatory variables or primitive truths*, and from these, they attempt to build up through (2) *formal operations or transformations*, increasingly more general or universal (3) *formal propositions or truth nets*. For all practical purposes these formal truth nets (linkages of propositions) may be regarded as the information output, or better yet, as the information content of a system.

The strengths of Leibnizian IS are the strengths that characterize all formal systems: consistency, rigor, logical coherence, precision, little or no ambiguity in the use of terms, conditions of proof, and so on. Their weaknesses are the weaknesses that beset all formal systems; for all their emphasis on logic, precision, and

222

rigor, Leibnizian IS are often extremely hard put to defend (except in the most vague and imprecise of terms) why they chose to solve a particular problem and why they chose to represent it in the manner used. For example, what justified their particular choice of primitive (why *are* they primitive?) explanatory or modeling variables. Leibnizian IS are also often accused of being empty; they are rich in analytic or formal content but extremely low in experiential or empirical content. A common example is the kind of operations research (OR) activity that places extreme emphasis on the building and exploration of the analytic consequences of very sophisticated mathematical models but places little emphasis on the development and use of equally sophisticated methods of data collection and analysis for getting the necessary input data to make the models operable. Given the almost complete preoccupation of OR with purely analytic systems, one can well understand why they have neglected this important side of their scientific development.

The case of OR illustrates another important point. Leibnizian activities are not just to be found in the philosophical literature, but throughout all of science. Indeed, the case of OR is particularly instructive because it vividly illustrates another characteristic associated with Leibnizian inquiry, the kinds of problem situations for which Leibnizian inquiry is most suited — the class of *well-structured problems*. Almost without exception, OR has been applied exclusively to problem situations that are well-structured in the sense that they admit of a well-defined analytic formulation as well as solution. Those problems that are ill structured and hence cannot be well formulated have either received scant attention in the OR literature [315] or have tended to be dismissed as either meaningless or insoluble. Perhaps even worse, OR has tended to regard every problem situation as Leibnizian, even where a mathematical approach to the problem was either not called for or highly suspect [70, 75].

It is important to clarify further the differences between well-structured and ill-structured problems. Although the terms could be defined in many ways, the difference between well-structured and ill-structured problems can be rather easily captured in terms of the class of decision problems. A decision theory problem can be defined as follows: to choose from among a set of acts — A_1, ..., A_m — that act — A_i — which optimizes (in some sense) the decision maker's (Z's) return U_{ij}, where U_{ij} is the utility or value to Z of the outcome O_{ij} corresponding to the doublet (A_i, S_j) where $\{S_j\}$ is the set of the "states of nature" [265].

There are three basic kinds of *structured* decision problems. A decision problem under certainty is one for which the sets $\{A_i\}$, $\{U_{ij}\}$, $\{O_j\}$, and $\{S_j\}$ are known. In addition, there is a known deterministic relationship holding between the choice of an A_i and the occurrence of an O_j. If the relationship between A_i and O_j is probabilistic and known (i.e., the probabilities P_{ij} are known), then we have a decision problem under risk. If the probabilities are not known (but only the probabilities are unknown), we have a problem under uncertainty. The first two kinds of structured problems are well structured because unambiguous rules exist for selecting an optimal course of action. The third kind of problem (uncertainty) is merely structured because an unambiguous rule does not exist for selecting an optimal course of action or act A_i.

An *ill-structured* or *wicked* decision problem is one for which one or more of the $\{A_i\}$, $\{U_{ij}\}$, $\{O_j\}$, and $\{S_j\}$ terms or sets is unknown or not known with any degree of *confidence*. Well-structured problems are problems about which enough is known so that problems can be formulated in ways that are susceptible to precise analytic methods of attack. The biggest problem connected with ill-structured problems is to define the nature of the problem. Ill-structured problems have an elusive quality that seems to defy precise methods of formulation. Most social problems seem to be of this kind.

Logic of Science

A significant part of the literature in the philosophy of science has been Leibnizian in execution if not in spirit. Where the philosophy of science has been Leibnizian in execution, it has been concerned with the direct construction of some Leibnizian model or explanation of the workings (or aspects) of science. The logic of science best typifies this concern. Where the philosophy of science has been Leibnizian in spirit or intent, it has argued that the ultimate explanation or understanding of science will be Leibnizian (i.e., logical), or that science is fundamentally Leibnizian in character. One of the best presentations of this point of view is to be found in Herbert Feigl's short but excellent paper, "The 'Orthodox' View of Theories: Remarks in Defense as well as Critique" [129]. Feigl's paper embodies many of the crucial distinctions central to Leibnizian inquirers; for example, on the point that theories are formal propositions built up from symbolic primitives he writes: "In fairly close accordance with the paradigm of Euclid's geometry, theories in the factual science have for a long time been viewed as hypothetico-deductive systems. That is to say that theories are sets of assumptions, containing primitive, i.e., undefined terms. The most important of these assumptions are lawlike, i.e., universal, propositions in their logical form" [129, pp. 4—5].

On the point that such systems are devoid of any empirical content Feigl writes: "Concepts thus defined are devoid of empirical content. One may well hesitate to speak of concepts here, since strictly speaking even logical meaning as understood by Frege and Russell is absent. Any postulate system if taken as (erstwhile) *empirically uninterpreted* merely establishes a network of symbols. The symbols are to be manipulated according to preassigned formation and transformation rules and their meanings are if one can speak of meanings here at all, purely formal" [129, p. 5].

And finally, Feigl writes on the point that such systems are not just "devoid" of any empirical content but that the realms of theory and of observation are completely disjoint. In the language of systems analysis, it is often said that the activities of theory-building and of data-gathering are supposed to be completely *separable from*[2] one another. "In view of the orthodox logical analysis of scientific theories it is generally held that the concepts (primitives) in the postulates, as well as the postulates themselves, can be given no more than a partial interpretation. This presupposes a sharp distinction between the language of observation (observational language; O.L.) and the language of theories (theoretical language; T.L.). It is asserted that the O.L. is fully understood. Indeed, in the view of Carnap, for example, the O.L. is not in any way theory-laden or contaminated with theoretical assumptions or presuppositions" [129, p. 7].

Lockean IS

Lockean IS are the archetypes of experiential, inductive, consensual systems. Lockean IS emphasize the purely sensory, empirical aspects of human knowledge. Where Leibnizian IS build up increasingly more general — formal — propositional truth nets from elementary — primitive — analytic truths, Lockean IS build up increasingly more universal inductive generalizations (fact nets) from the elementary sensory data of raw experience. Where in the Leibnizian IS it is a set of formal operations that transforms the primitive elements into elements of information, in the Lockean inquirer it is the function of human judgment which accomplishes the transformation from raw data to factual information. The particular aspect of human judgment which most often accomplishes this is the function of agreement. Thus, for example, two or more observations or observers are judged objective if they are in sufficient agreement with one another. A beautiful example of a modern Lockean inquirer is that of a Delphi exercise or Delphi policy advisor [448]. In a Delphi, the raw data inputs are the human participants who take part in the exercise, i.e., their raw opinions or initial judgments on some issue. It is the degree of agreement among the group of Delphi participants that transforms their raw judgments into factual information or into the well-substantiated policy of the group. Indeed, agreement is so strong a methodological or epistemic principle that in the large majority of Delphis conducted to date, a Delphi is terminated when the level of agreement reaches some satisfactory level. The resulting judgments that survive the group agreement process become the empirical information output or content of the system. Agreement is so important a principle that it can even serve to eliminate those participants whose judgments do not sufficiently agree with those of the surrounding Lockean community from round to round.

The presumed importance of agreement for science is attested to by the large numbers of scientists and philosophers of science who have repeatedly stressed its crucial role. For instance, Ziman's *Public Knowledge* [486] is an interesting and vigorously argued consensual view of science. Ziman persistently argues the merits of *consensus*, both as a method and as a goal of science: "The objective of science is not just to acquire information nor to utter all noncontradictory notions; its goal is a *consensus* of rational opinion over the widest possible field" [486, p. 9]. And, "The argument is that science is unique in striving for, and insisting on, a consensus" [486, p. 13].

Consider the thoughts of Bentley Glass:

> The objectivity of science depends wholly upon the ability of different observers to *agree* [emphasis added] about their data and their processes of thought... In the last analysis *science is the common fund of agreement between individual interpretations of nature* [emphasis added]. What science has done is to refine and extend the methods of attaining agreement...

226

All that can be claimed for science is that it focuses upon those *primary observations* [emphasis added] about which human observers (most of them) can *agree* [emphasis added] and that it emphasizes those methods of reasoning which, from empirical results or the successful fulfillment of predictions, most often lead to mental constructs and conceptual schemes that satisfy all the requirements of the known phenomena [151, p. 1256].

The strength of Lockean IS lies in their ability to sweep in rich sources of experiential data. In general, the sources are so rich that they literally overwhelm the current analytical capabilities of most Leibnizian systems. The weaknesses, on the other hand, are those that beset all empirical systems. Although experience is undoubtedly a rich source of knowledge, it can also be extremely fallible and misleading. Further, the simple sensations, facts, or observables of the empiricist have always, on deeper analysis, proved to be exceedingly complex and further divisible into other entities themselves thought to be indivisible or simple, ad infinitum. It can be shown that it is impossible for Lockean inquirers to have even simple sensations or experiences, unless they have already had some powerful generalizations or properties built into them that would allow them to recognize or have simple sensations. In science, it is impossible for an observer to observe raw data without any prior theoretical notion on his part about what is important to observe as raw data, or at the very least, what shall count as raw data. Further, if Lockean inquirers are ever to induce some general laws, it is only because they have already presupposed some general a priori laws in their design that makes (Kantian) induction, possible. (See also note 4 on these points.)

No less troublesome in the design of Lockean IS is their almost extreme and unreflective reliance on agreement as a principle for producing knowledge or information out of raw data. The costs of agreement can become too prohibitive, and agreement itself can become too imposing. This is not to say that agreement has nothing to recommend it. It is just that agreement is merely one out of many epistemic rules for producing knowledge out of experiential data. If we are to believe Kuhn [247, 248], much of the agreement in science, rather than being freely achieved or "naturally arising out of the nature of things," is instead the result of the scientist's mode of education. If Kuhn is correct, the scientist's mode of education is such that it reinforces him to look for and emphasize the agreement aspects of nature rather than the disagreement aspects.[3] There is a growing body of literature in the psychology of science that indicates that science has tended to attract disproportionately those personalities who have a compulsive need for agreement [203—205, 286, 456]. The point is that agreement in science is as much a function of our need for it, and hence our nature, as it is of the nature of things themselves.

The danger with agreement is that it may stifle conflict and debate when they are needed most. As a result, Lockean IS are best suited for working on well-structured problem situations for which there exists a strong consensual position on the nature of the problem situation. If a strong consensual position does not exist, or if the consensual position is suspect no matter how strong it might be, Kantian and Hegelian IS may be called for.

Kantian IS

Kantian IS are the archetype of synthetic multimodel systems. Kantian IS emphasize both the formal and the experiental — the integrative — aspects of human thought. They are synthetic in the sense that they attempt to reconcile (synthesize) the demands of philosophic rationalism (Leibnizian IS) and that of philosophic empiricism (Lockean IS). They are multimodel in the sense that, where Leibnizian and Lockean IS usually build only one formal model or only one inductive generalization, Kantian inquirers produce at least two alternate models, either of which will equally fit the data or explain the primitive explanatory variables. For those who identify the essence of Kantian thought with a single set of fixed categories, this particular feature of the characterization of Kantian IS deserves special mention. I believe that one of the most fundamental features of Kantian inquiry is the insight that, in order to engage in the process of observation, one must make certain assumptions a priori about the nature of the world (or in the language of science, have available certain prior theoretical notions) that will make the act of observation possible.[4] In Kant's time the range of assumptions was greatly constrained. Thus, he could assume or prove the necessity of but one geometry (Euclidean), one physics (Newtonian), and one logic (Aristotelian). Given the advance of science since Kant's time, the range of available assumptions open to us is greatly expanded. As a result, we are able to give multimodel explanations of phenomena. We can now pose the interesting question of how phenomena or problem char-

228

acterizations change as we vary their representation across category sets.

Kantian IS invoke at least two Leibnizian submodels that give alternate, and in many cases, even contradictory explanations of the same phenomenon or problem. Each submodel directs us to collect a different set of supporting data on the same phenomenon, each data set being matched to the guiding submodel. The hope is that, out of the multitude of models plus supporting data sets produced, there will be one that will be best or optimal for a decision maker's problem. Instead of presenting a decision maker with only one model or information set for his final approval or disapproval, Kantian IS explicitly give the decision maker the unique opportunity to see how his problem changes as the models of it change.

Social problems especially seem to necessitate a Kantian approach. For example, alcoholism and drug abuse are neither public health, medical, psychiatric, or social welfare problems exclusively. They are all of these and none of them. Alcoholism and drug abuse are multidisciplinary problems. They are the sole or exclusive province of no one single discipline. Indeed, as the design of our cities, transportation systems, etc. show, the attempt to regard the solution (or the gathering of information) on important social problems as the exclusive concern of one discipline or point of view has been nearly fatal.

The strength of a Kantian IS is that it counters the weaknesses of both Leibnizian and Lockean IS. The weaknesses are: (1) there is no guarantee that the multiple models will include the right model; (2) there is the danger that the decision maker will be more overwhelmed than aided by the multitude of models; and (3) Kantian IS are more costly to operate — with the addition of each model, the cost and time of the system goes up disproportionately. Kantian IS are best suited for handling problems of moderate ill-structure where the alternate views of problems are absolutely essential for problem definition. For a view of science that has strong Kantian elements, see Rose [378].

Hegelian or Dialectical IS

Hegelian or dialectical IS are the archetype of conflictual, synthetic systems. Hegelian IS emphasize the antagonistic and the antithetical, the conflictual aspects of human thought. An Hegelian inquirer is designed to present the strongest possible debate on any issue. In a Kantian IS the alternate submodels are not necessarily antagonistic. They may in fact be highly complementary. There may be a great deal of overlap between them. They may merely represent different complementary ways of looking at the same issue. In an Hegelian inquirer, the overlap (or in set-theoretic terms, the *intersection*) is zero. In an Hegelian inquirer, the submodels (of which there are at least two) are in complete opposition on almost any and every point. As anyone familiar with Hegelian thought knows, the conflict runs so deep that the two opposing points of view are the deadly enemies of one another [177, 239].

On any problem, an Hegelian IS will build at least two completely antithetical representations. Hegelian IS start with either the prior existence (identification), or the creation of, two strongly opposing (contrary) models of a problem. These opposing representations constitute the contrary underlying assumptions regarding the theoretical nature of the problem. Both of these representations are then applied to the same Lockean data set in order to demonstrate the crucial nature of the underlying theoretical assumptions, namely that the same data set can be used to support either theoretical model. The point is that data are not information; information is that which results from the interpretation of data [16, 72]. It is hoped that out of a dialectical confrontation between opposing interpretations (e.g., the opposing expert views of a situation [282]), the underlying assumptions of both models (or opposing policy experts) will be brought up to the surface for conscious examination by the decision maker who is dependent on his experts for advice. It is also hoped that, as a result of witnessing the dialectical confrontation between experts or models, the decision maker will be in a better position to form his own view (build his own model or become his own expert) on the problem that is a creative synthesis of the opposing views. (For an actual case study of this process in strategic planning, see Mason [282]; for a theoretical model of a dialectical IS, see [304, 305]; for a treatment of decision theory from a dialectical point of view, see Mitroff and Betz [315].)

Where in the Lockean IS the epistemic rule guaranteeing the process is agreement, in the Hegelian IS it is intense conflict — the presumption that conflict will expose the assumptions underlying an expert's point of view that are often obscured precisely because of the agreement between experts. For this reason, Hegelian

230

IS are best suited for wickedly ill-structured problems. These are the problems that, because of their ill structure, will involve intense debate over the true nature of the problem. Conversely, they are extremely ill suited for well-structured, clear-cut problems because here conflict may be a time-consuming nuisance.

It is extremely important to appreciate the differences between this application of dialectical thinking and the more traditional forms. The big difference is that the new has been formulated in such a way as to make it susceptible to scientific investigation. Behavioral experiments, utilizing published models [304, 306], have been undertaken to test the efficiency of the dialectical IS in presenting policy issues. It is also important to appreciate how the dialectical inquirer differs from ordinary debates as well as from most of the social science research that has been conducted utilizing conflict or persuasion as a means for effecting attitude change [87, 191, 202, 210, 233, 266]. In an ordinary debate, or in the majority of research conducted to date on attitude change, both sides are free to introduce whatever data they want as well as whatever arguments they desire. The arguments are not in the form of strict contraries, and both sides are not constrained to apply their arguments to the same data set. As a result, an ordinary debate often confuses the two and makes it extremely difficult for the listener or decision maker to sort them out. In a dialectic debate, the two are introduced separately. First, the opposing theoretical policy assumptions dividing the two experts are laid out side by side, perhaps on a display board, by the designer of the debate. Second, each theoretical policy assumption is then separately applied to the data-set in order to draw out the implications of that individual policy assumption. This, of course, introduces an element of artificiality that real debates do not have, but it also introduces an element of clarity and makes possible a more systematic investigation of the effects of individual policy assumptions. Any experimental investigation, particularly with a new tool, is always a trade-off between reality and clarity.

The strengths of an Hegelian inquirer are: (1) The decision maker is actively involved in the information creation process; unlike some of the other inquirers, the user of the information is no longer a passive recepticle for the end-product but an active creator of what he will use. (2) Unlike most scientific information systems that speak in the dry language of facts and symbols, Hegelian IS speak in the more active language of drama [72, 282, 316]. They use conflict both as an intellectual and as a dramatic device to induce the decision maker to take an active interest in the system. The weakness of Hegelian IS is that they may create conflict when it is not there or appropriate. Further, not all decision makers can either tolerate conflict or learn from it optimally. And they are costly and time consuming because they involve the creation and use of at least two equally credible theories or experts.

For an extremely interesting view of science that has strong conflict elements, see Feyerabend [132—135]. See Churchman [72] for a systematic exposition of the properties of Hegelian IS.

Singerian—Churchmanian IS

Singerian-Churchmanian IS are the archetype of synthetic, interdisciplinary holistic systems. Singerian IS emphasize the richly diverse modes of human thought: the scientific, the ethical, and the esthetic. Singerian IS are unique among philosophic-scientific inquirers in the sense that they are among the few systems that attempt to integrate within one frame of reference the poles of human thought that are so often at odds with one another. Because they are the richest of all the inquirers encountered thus far, they are also the most difficult to describe. Because of this richness, only a few selective features can be described here. What each of the foregoing inquirers takes as a primitive variable requiring no further explanation, Singerian inquirers take as objects for infinite study. Thus, formalism, experience, agreement, synthesis, multiplicity and disagreement are all topics for deeper study by some other inquiring system process [72]. (For example, for a Leibnizian analysis of a Lockean variable such as agreement, see [241]).

One extremely important aspect of Singerian inquirers is how they attempt to reconcile the scientist's and the ethicist's notions of information. This has to do with perhaps the most distinctive feature of Singerian inquirers, their language. Singerian inquirers speak almost exclusively in the language of commands or instructions. The point is a subtle one. With the exception of the Kantian and Hegelian inquirers, Leibnizian and Lockean IS make it appear that their representations of reality are synonymous with reality (e.g., "The structure of reality is logical"). Singerian inquirers, on the other hand, recognize that all descriptions are only representations and that to a large extent their reality only comes about if

232

each inquirer can convince enough people (decision makers) to regard the description as real. In effect, what each inquirer does (and this is also true of every scientific model) is to issue an implicit command — e.g., "So regard the world that the way my model represents it is true!" Thus, none of the models of science are to be taken as true or real in the literal sense. Rather their proponents implore us to take them as true in very subtle ways. Notice that under this conception of science, we do not accept a model just because it is supported by empirical data, although this is important. Rather, we accept a set of data as relevant to the support or nonsupport of a model because of the prior acceptance of the model as an intelligible way of representing or interpreting nature. If we accept the pleas and commands of their proponents, then we will be able to share concepts, and then we will be able to perform the same experiments, and as a result we will uncover the same supporting data, and interpret phenomena in the same light. In this way, a model which is the scientific expression of a command basis and which is supported by the right data and arguments of enough reputable and influential scientists, will come to be taken as real. In time, the model and the basis can even come to be synonymous with the fundamental structure of nature itself — i.e., be regarded as a natural law of science.

Singerian inquirers make this command basis explicit, so that it, too, can become a topic for explicit inquiry, and ultimately we can choose explicitly between different command bases. As Churchman has put it:

> In place of "X is P" [i.e., the language of is], Singer therefore suggests something like "The object observed is to be taken as having property P plus or minus E." The "is to be taken" is a self-imposed imperative of the community. Taken in the context of the whole Singerian theory of inquiry and progress, the imperative has the status of an ethical judgment. That is, the community judges that to accept its instruction is to bring about a suitable tactic or strategy in the grand teleological scheme. The acceptance may lead to social actions outside of inquiry, or to new kinds of inquiry, or whatever. Part of the community's judgment is concerned with the appropriateness of these actions from an ethical point of view. Hence, the linguistic puzzle which bothered some empiricists as to how the inquiring system can pass linguistically from "is" statements to "ought" statements is no puzzle at all in the Singerian inquirer: the inquiring system speaks exclusively in the "ought," the "is" being only a convenient *facon de parler* when one wants to block out uncertainty in the discourse. As a computer programmer would say, the whole design is instructions, including the "data base." [72, p. 202].

233

Because we are required to make strong ethical judgments in order to determine and validate all of the so-called purely factual statements of science (e.g., "x is p" or "water is H_2O"), for Singerian inquirers there is no problem of how to get ethics into every scientific judgment. It has been there all along [257, 307, 362]. Again, as Churchman has written: "Positivists have told us that we cannot derive the 'ought' from the 'is.' This may be true, but it is irrelevant: we never are basically in an 'is' language in the first place" [73, p. 115]. The real problem is how to make the ethical explicit and how to go about deciding which ethical judgments we wish to incorporate. In a Singerian inquirer, information is thus no longer merely scientific (formal, experiential, or synthetic) but ethical as well. A Singerian Inquiring System, in other words, attempts to point out the ethical implications underlying a proposed set of technical actions.

Although Singerian inquirers also rely on agreements — for example enough people who accept a similar command basis — the sense of agreement is very different from that which is incorporated in the Lockean inquirer. In the Lockean inquirer, raw data are received which, via agreement, are then transformed into information or objective knowledge. Supposedly the raw data are themselves not the result of particular kinds of prior agreements or epistemic judgments. From a Singerian point of view, however, Lockean inquirers only start with raw data because they have already agreed implicitly with respect to this conception of the origin and growth of knowledge. From a Singerian point of view, the Lockean notion that one starts with raw data is itself the result of a prior epistemic decision. The Lockean representation is the result of prior agreements, implicit to be sure, on the nature of reality. Some other features of Churchmanian—Singerian inquiry, namely, its interdisciplinary and antireductionistic basis, were described in Chapter One.

Table 18 summarizes some of the more salient features of these inquiry systems and considers the possibility of using each of these IS as complex decision-making stimuli and responses. In everyday life the stimuli (Ss) and responses (Rs) we face are exceedingly complex. We rarely encounter the simple physical entities of the behavioral scientist's laboratory, especially in isolation from other entities or systems. More often than not, the most persistant stimuli that we are continually forced to respond to are complex entire *systems* of linguistic and conceptual stimuli which demand equally complex patterns — i.e., systems — as responses. It is im-

Table 18.

Inquiring systems

IS	As stimuli — A subject is presented with	As responses — A subject is required to respond with or to produce
Leibnizian		(1) a set of primitive (undefined) formal elements (symbols);[5] (2) a set of explicit, formal rules or operators for forming ... (3) a model or a single set of complex relations (propositions) that can be ... (4) explicitly shown to follow from (1) by means of (2).
Lockean		(1) a set of primitive (undefined) experimental elements (sensory qualities, elementary observations, or raw data);[5] (2) a set of agreement producing experiential operators (individual perception, expert judgment, group decision-making, T-groups) for forming ... (3) a consensual position or a single set (linkage) of factual propositions that can be ... (4) explicitly shown to follow from (1) by means of (2).
Kantian		(1) at least two alternate (complementary, but potentially divergent) sets of Leibnizian elements and operators);[6] (2) a set or sets of Lockean primitive, experiential elements such that ... (3) when each of the Leibnizian element and operator sets is applied to the Lockean element set(s), a set of multiple Lockean—Leibnizian (integrative) models (interpretations) is produced such that ... (4) the interdependence between the Leibnizian and the Lockean elements is explicitly demonstrated and the multiple interpretations are made available for explicit consideration (choice).
Hegelian or Dialectical		(1) at least two completely antithetical (strictly contrary) sets of IS elements (differing fundamental assumptions) and operators such that when they are applied to ... (2) a common Lockean set of primitive, experiential elements (a data set), ... (3) the underlying IS model assumptions are brought up to the surface for a decision maker's conscious inspection for the purpose of ... (4) allowing (aiding) the decision maker to form his own model assumptions that are a synthesis of the antithetical IS model assumptions.

Table 18 (continued)

IS	As stimuli — A subject is presented with	As responses — A subject is required to respond with or to produce
Singerian—Churchmanian	(1) a set of Leibnizian, Lockean, Kantian and Hegelian models such that ...	
	(2) any model may be recursively applied to any other model (including itself) for the purpose of ...	
	(3) elucidating the distinctive, characteristic assumptions underlying each model and the entire inquiry process so that ...	
	(4) the more nearly strictly scientific (technical) features of inquiry (Leibnizian, Lockean, and Kantian) can be integrated with the ethical features of inquiry (Hegelian/Churchmanian).[7]	

portant to emphasize that the summary represents the most complex and complete form that each of these IS as Ss and Rs can take. Not every experiment will or need embody every defining characteristic of these inquiry systems.

Taxonomy of IS experiments

Combining stimuli and responses in all possible ways results in 25 archetypal kinds of potential experiments. To date the behavioral sciences have systematically explored but a very small subset (mainly, the upper left-hand corner of Leibnizian, Lockean S-R combinations in Table 19) of the total kinds of experiments that are generally available. Table 19 indicates briefly the kinds of problem situations for which the IS experimental combinations would seem most appropriate for study.

A brief elaboration of a few of the cells in Table 19 is in order. The Leibnizian/Kantian and Lockean/Kantian cells involve the transformation of a single system (formal or experiential) into a set of multiple systems (Kantian). The Leibnizian/Hegelian, Lockean/Hegelian, and Kantian/Hegelian cells involve either the transformation of single systems (Leibnizian/Lockean) or a set of complementary multiple systems (Kantian) into a set of strongly conflicting systems (Hegelian). The Hegelian/Leibnizian and Hegelian/Lockean cells involve the transformation of a set of strongly conflicting systems (Hegelian) into a single, well-defined

236

Table 19.

Taxonomy of IS Experimental Problem (Decision-making) Situations

IS as stimuli ＼ IS as responses	Leibnizian	Lockean	Kantian	Hegelian	Singerian
Leibnizian	Traditional experimental design: formal testing of well-defined experimental hypotheses		Idea generation: production of possibilities hypothesis-production; brainstorming	Conflict Generation and/or production	Derivation of the ethical implications of scientific/technical ideas (models)
Lockean					
Kantian	Detailed development or analysis of multiple ideas/hypotheses		Alternate futures: elaborative idea generation, elaborative production of possibilities	Conflict production	Ethical implications of futuristic planning
Hegelian	Conflict resolution or suppression		Conflict management		Conflictual futuristic planning
Singerian	Conversion of scientific/ethical ideas into scientific/technical models			Conflict perpetuation	I^2, or inquiry into the process of inquiry: modeling of the art of building models

system (Leibnizian or Lockean). The Hegelian/Kantian cell involves the attempt to manage conflict by giving a Hegelian stimulus a number of alternate (but not strongly conflicting) interpretations so that the decision maker or subject of the experiment can choose between the interpretations — i.e., he can choose the one most suitable for his problem. With these ideas as background, we can turn again to the notion of objectivity.

Toward a Combined Social Psychological and Philosophic Theory

Churchman wrote, "no observation can become objective unless the observer is also observed objectively" [72, p. 150]. Leibnizian and Lockean notions of objectivity are the most common and the ones that most widely pervade science. Leibnizian notions range all the way from (1) the use of explicit well-formed rules to deduce systematically the conclusions (theorems, logical truths) of formal axiomatic systems, to (2) the investigation of formal systems for the confirmation and falsification of provisional scientific truths or hypotheses, to (3) formal systems (rules) for the design, conduct, and evaluation of scientific experiments. Although there is thus no single notion that encompasses all the properties of all the various Leibnizian notions, there is nevertheless a common thread that runs throughout all of them. Whatever the particular notion, the accent is on the depersonalized, careful testing and scrutiny of scientific knowledge by impersonal, well-specified, formal tests. The objective character of scientific knowledge is founded on the impersonal character of the rational test procedures that a scientist uses to muster evidence either in support of or in denial of his purported claims to discovery and to knowledge. The emphasis, in other words, is on the character and testing of that which is discovered and not on the behavioral and historical characteristics of the discoverer.[8] As an attitude, this approach to objectivity is obviously cognitive or cerebral. In the discovery phase of scientific knowledge, the strong emotions and attitudes a scientist displays are perceived as irrelevant; in the testing phase, strong emotions are perceived as a contaminant to be controlled if not eliminated altogether. Apparently the only approved attitude in the testing phase is one of impartiality and detachment.[9]

Lockean conceptions also emphasize the impersonal character of scientific objectivity. But whereas in the Leibnizian approach the emphasis is typically on the formal, rational test nature of

238

scientific objectivity, in the Lockean approach the emphasis is on the observational character of scientific knowledge. The core of most Lockean positions hinges on the idea of consensus among the observational reports of different competent observers of the same phenomenon. In other words, Lockean positions, at some point or another, rest on the assumption that all competent observers in the same experimental or experiential circumstances should be able to make the same observations. The following quote from the psychologist S.S. Stevens is a typical statement of this position: "'Objectivity' in science is attained only when facts can be regarded as independent of the observer; for science deals only with those aspects of nature which *all normal* [emphasis added] men can observe alike" [422, p. 327].[10] Here again the underlying attitude is cognitive, for the emphasis is on such nonaffective notions and attributes as neutral observations, hard facts, and the idea of the neutral, disinterested observer.

When we move beyond some of the more prevalent Leibnizian and Lockean notions, the explication of alternate IS notions of objectivity becomes considerably more difficult because they become considerably more sophisticated and more difficult to understand. To be sure, the traditional Kantian notion is clear enough: objectivity is the subsumption of individual experience (the given in sensuous intuition) under the general categories of the understanding which give the individual experience its universal or necessary character and thus render the experience the same for all men. However, the interest here is not so much with the classic Kantian notion of objectivity as with a more modern interpretation that takes its basic inspiration from the original Kantian idea.

By now we should have learned at least this much from Kant: (a) every set of scientific observations is dependent on the presumption of some theory for its collection, and (b) in order for any set of observations "to test a theory$_1$" we must presuppose (or demonstrate) some theory$_2$ as to how the observations relate to a test of the theory$_1$. In this sense, the testing of a theory by observations is anything but direct [314].

The case becomes even more tortuous if we recall one of the attitudinal items from Chapter Five. The item is, in fact, almost a direct quote from Churchman: "Theories of science are not tested by a set of data [or observations] arrived at independently of the theory itself [77, p. 133]." That is, one can show that, for a set of observations to test a theory, it must in some fashion either be a function of or related to that theory.

239

The point is one that has continually been made by Feyerabend in a series of provocative papers [132—135]. If all observation is theory laden — as it cannot help but be if theory always guides the determination of what shall be regarded as relevant to the testing of the theory — then the explicit development of only one theory on any phenomenon (no matter how well confirmed that theory may be) can be detrimental to the continued testing of that theory. The adoption of only one theory on any phenomenon can have the effect of uncovering only one set of supporting or disconfirming data. As long as we restrict ourselves to one theoretical base, we cannot know whether we have found the strongest supporting or disconfirming data set. The question does not even make sense as long as we confine ourselves to just one theoretical base. Thus, Feyerabend proposes that in order to test a theory we must continually produce strongly competing alternatives to that theory. Further, Feyerabend contends, this will always be the case. The need for producing alternatives is not something that one eventually outgrows. Precisely because we can no more fully validate a theory than we can fully disconfirm it, we must of necessity always be committed to a theoretical or methodological pluralism.

The Feyerabendian notion of theory testing leads to an alternate conception of objectivity to the Leibnizian and Lockean notions we have been pursuing. If I understand the implications of Feyerabend's ideas, the idea of thrusting opposing theories against one another not only leads to a test of one or both but, just as significantly, it exposes the theoretical assumptions and intellectual commitments which underlie each theory. The process of opposition reveals what each theory requires that we presuppose about the nature of the world (the theory's field of application) so that experiments can be designed to collect data on that world. In terms of this conception of inquiry — for the point is that objectivity is an integral part of a conception of inquiry — objectivity is not merely concerned with the testing of scientific ideas. It is just as concerned with the systematic examination of the theoretical commitments that shape those ideas. To put it differently, objectivity is not just confined to or a property of the end phase (testing) or end products of scientific knowledge (theories, laws). It is a property of the whole process of doing science. This is why the logician's distinction between the contexts of discovery and justification is so distressing. It not only promotes a concept (understanding) of science but a concept (understanding) of objectivity

240

that is limited to the end phase of scientific work. By this I do not mean to imply that the Leibnizian and Lockean notions are not important, for they are. I do mean to say that they may not be concerned with the most important conception of objectivity — the scientist's need to be able to inspect how his presuppositions influence his total performance, from the initial conception of his ideas to their final testing. It is too limiting and confining to relegate objectivity to the testing of the end products of a scientific study, when in reality one presupposes a notion of objectivity that guides the entire process. One would certainly like to have a conception of objectivity that begins to make possible the evaluation of the entire research experience, from its initial conception to its end.

In one sense, an Hegelian IS notion of objectivity differs little from the immediately preceding Kantian notion. Objectivity consists of the ability to inspect and to understand how one's underlying presuppositions (which in the Hegelian IS become akin to a world view) shape one's interpretations of the world. In the broader and perhaps more traditional sense of Hegelian inquiry, objectivity is that new world view which (1) survives or is the product of an intense debate, and (2) is able to integrate (synthesize) on a higher level the prior world views of which the new view is a result.

This hardly begins to exhaust the matter, because there are a host of other related Hegelian aspects to objectivity. For example, one of the most intriguing Hegelian ideas is that objectivity is not a concrete something (e.g., a test) that a single observer or mind (or in Hegel, a world process) can determine of itself or its products. Instead, objectivity is a process whereby one mind actively observes and interrogates another. In effect, every mind is too close to itself to be able to observe and determine accurately its true states, particularly by definition its unconscious states. For instance, there are times when other minds can more accurately read the states of my mind — e.g., my underlying motives, than I can myself [313]. Under this conception, objectivity becomes tied to the ability of one mind to observe and study the states of another mind. As Churchman has put it, "Objectivity is a property of an observer of a subject, i.e., a property of self-reflection [72, p. 158]." There of course must be more to this conception than the mere idea of one mind observing another. If that is all there were to the idea, we would be back to the Lockean notion of objectivity: Objectivity is that which is grounded in the agreement among the observations of different observers. The distinguishing mark is that, in the Hegelian notion of one mind (the observer) observing another (the observed), the observer and the observed become opposed to one another; they disagree rather than agree over how the observer characterizes the observed, and an ensuing dialectical debate takes place of the properties to be assigned to the mind of the observed:

> The essence of "mechanical" observation is alienation: the observed subject is opposed to the observer. Either the subject is passive and the observer active, or else the observer receives "inputs" (and hence is passive) while the subject creates "outputs" (and hence is active). *The observed and the observer cannot be the same mind, and must be two opposing aspects of a process* [emphasis added]. The alienation is well known in experiments in which humans (or other living beings) are subjects, or in interviews in which the behavior and attitudes of people are being studied. *The experimenter or interviewer is the observer and is a different kind of person from the subjects* [emphasis added]. He is supposed to have no prejudices, to be rational, to be completely honest in his reporting, not to care who is right, and so on, while the subjects are interesting only because they have prejudices, are irrational, dishonest, self-seeking, etc.
>
> The observer-of-the-subject is not a dispassionate "other mind"; it is passionately dedicated to destruction of the subject's conviction. The very activity of observing the subject creates a mood of opposition [72, pp. 159, 173].

Whatever the means of resolving different perceptions of the same thing, the Hegelian idea is that objectivity is not a formal property or function of either the thing observed (the object) or of the observer, but that objectivity is a function of the interplay between observer and observed (object). Whereas the Lockean IS uses the decision-theoretic or epistemic rule of *agreement* to resolve the differences between observer and observed, the Hegelian IS depends on and utilizes the notion of *opposition* between observer and observed to explicate the meanings of the observations as well as the nature of the observer and observed. This notion of objectivity would seem especially suited to the behavioral sciences, where more often than not the objects of our investigation are our fellow men, if not the class of living beings in general.

The Churchmanian or Singerian notion of objectivity may be the most difficult to convey precisely because it tries to utilize each of the preceding notions to build a comprehensive notion of objectivity. In explicating the Singerian idea, it is easiest to start with the Hegelian idea of one mind, the observer, observing another mind, the observed. In brief, the Singerian idea consists of a

242

more systematic expansion of the ways that different observers can observe or study one another. In effect, objectivity in the Singerian sense is a further generalization of the Hegelian idea of an observer of the observed.

One of the purposes of the observer (O_2) of the observed (O_1) is to explicate the presuppositions of the observed. But because the observer (O_2) has presuppositions that affect his observations and characterizations of the presuppositions or general behavior of the observed (O_1), the Singerian suggestion is to add another observer (O_3), the observer (O_3) of the observer (O_2), whose function is to explicate the presuppositions of the observer (O_2). The Singerian notion of objectivity consists of an explicit paradigm whereby we can study how the observer's (O_2) presuppositions affect his study (understanding) of the presuppositions of the observed (O_1). Of course, there is no end to this process of an observer of an observer of an observer, so theoretically it could be carried on indefinitely. Thus, there are no absolute reasons for stopping the process with an additional observer (O_3) of an observer (O_2). However, the purpose of the paradigm is not to include all the possible terms of the process or series, but rather to include enough terms so that the basic idea can be conceptualized and the phenomenon can then become a legitimate object of study. If one would study the effect of the observer's (O_2) presuppositions on the characterization of the observed (O_1), one has to include, at a very minimum, the basic notion of an observer (O_3) of an observer (O_2) that would make the study of the phenomenon possible. This can be put another way. It has now been realized through the studies of Rosenthal [379] and Orne [336] that the scientific observer (O_2) or investigator is one of the most important variables in the design and execution of a scientific study, and as such the effect of the investigator (O_2) deserves systematic study. In these studies, Rosenthal and his colleagues have in effect served the role of O_3.

My close friend and colleague Frederick Betz has developed the notion in much greater detail [personal communication]. Betz considers systematically the triad, O_3 o_1 O_2 o_2 O_1. To O_3, Betz gives the name the *reflective observer* or the *other observing mind*. (See note 11.) Betz notes that this role has been traditionally fulfilled by the philosopher of science, whose primary concern has been to study the methods of the scientist. To O_2, Betz gives the name *observer* or *primary observing mind*. Betz notes that this role has been traditionally fulfilled by the scientist. To O_1, Betz gives the name *object*, noting that the object may either be a physical or a behavioral object of investigation. If the object is the subject of a physical investigation, then it will be a physical object; if it is the subject of a behavioral investigation, then it will be an animate subject. In either case, it is the object of an observer's observations. The symbols o_1 and o_2 refer to the possible modes of observation that O_3 and O_2 can choose from to utilize in their observations of O_2 and O_1 respectively.

Betz's next step is to consider all possible permutations of these three basic roles and to explicate what the permutations signify. Thus, for example P o_1 S o_2 O signifies the philosophy of science, i.e., the study or scrutiny by a philosopher P of the behavior (methods) of a scientist S engaged in some study of an object O. S o_1 S o_2 O signifies the science of science or S^2, e.g., the study of one scientist studying the behavior of another scientist studying some object. As a combined philosophic and social psychological investigation, this study is a combination of these two types — i.e., of S o_1 S o_2 O and P o_1 S o_2 O.

S o_1 P o_2 O signifies the scientific study of philosophy — e.g., the scientific study of the ways (and reasons for them) that philosophers typically study (characterize) the objects of the physical or behavioral world. In a similar fashion one can go through and assign meanings (labels) to each of the various triads, with each triad representing a significant kind of scholarly activity or tradition.

Because any one of the three observer or object positions (roles) can be fulfilled by a philosopher, a scientist, or an object (subject), there arises the possibility of $3 \times 3 \times 3$ or 27 distinct triads. If, in addition, we note that the o_2 and o_1 operators or modes of observing may be fulfilled by any one of the 5 IS (Leibnizian, Lockean, Kantian, Hegelian, and Singerian) — because in effect each of the IS specifies an archetypal way of engaging in a study of an object — then there arises the possibility of $3 \times 5 \times 3 \times 5 \times 3$ or 675 different possible modes of inquiry. Although such a schema or taxonomy can be used for many purposes, one of its prime purposes for this discussion is to indicate the different number of potential ways that one can engage in inquiry versus the actual number of limited ways that we currently engage in inquiry. Thus, for example, current work in the philosophy of science has tended almost exclusively to pursue one of two forms, P o_2 S o_1 O or P o_2 P o_1 S (a philosopher studying another philosopher's characterization of science). In addition, major emphasis has been devoted

244

to Leibnizian (formal) modes of characterization. This means that o_2 in P o_2 S o_1 O and o_1 in P o_2 P o_1 S have tended to be Leibnizian. The general schema attempts to point out that a whole host of unexamined and undeveloped other ways of doing philosophy of science remain. For example, S o_2 P o_1 S and O o_2 P o_1 S are but two of the other forms, and the o_2 and o_1 operators may take on any one of 5 possible modes.

The schema also helps to point out that, although they are important components, the Leibnizian and Lockean IS and their respective conceptions of objectivity are only components and not the whole of a systematic theory of inquiry or of a comprehensive theory of objectivity. From the standpoint of each of the individual triads, O_3 o_1 O_2 o_2 O_1, objectivity is not a property of the individual components of the triads; it is not a property of the different observers or of the different modes of inquiry used for observation. Rather, objectivity is a property only of each of the triads as a whole. If objectivity is a function of the ability to scrutinize critically one's unexamined and underlying presuppositions, then it is only in the triads as a unit that we achieve the systematic examination of an observer's presuppositions. This does not mean that every scientific study (S o_2 O) has to formally include an observer of an observer in order to qualify as scientifically objective. Rather, it means that the eventual establishment of a study's objectivity is potentially tied to the ultimate review of that study's assumptions by some other party. To a limited extent, this of course is the function now served by reviewers who scrutinize a study for possible publication. However, a more conscious awareness and utilization of the schema would make it even clearer that the reviewers are a fundamental part of the establishment of the study's objectivity. And so their choice and the design of their functions should be taken more seriously and given the more careful consideration that they so rightly deserve. For one, we might undertake even more systematic inquiries into the design and inspection of the crucial role of reviewers in science [108].

Even further, from the standpoint of the schema as a whole, objectivity in the broader sense is not a property of any one of the individual triads. It is a property of the schema as a whole, in the sense that the schema outlines a systematic way of investigating the presuppositions of any single triad. If the individual triads are an important way of conceptualizing the investigation of the presuppositions of any one of the individual components, then the schema as a whole is a way of conceptualizing the investigation of the presuppositions of any one of the individual triads. In this sense, the schema as a whole can be conceived of as the generalized observer (O_4) of the observer (O_3) of the observer (O_2). The schema directs us to ask such questions as: Would our results be the same if we adopted some other mode of inquiry, i.e., some other triad? What reasons can we give to justify the adoption of a particular triad for a particular kind of investigation?

In short, objectivity is ultimately related to our ability to outline a systematic plan for engaging in an inquiry into the nature of the assumptions underlying the general nature of inquiry. This is why objectivity is such a difficult and elusive concept to deal with. It is not merely a function of our current conceptions and practices of science. It is fundamentally related to our plans for expanding the scope of inquiry, so that we can inquire into the presuppositions behind our current conceptions of inquiry. For this and many other reasons beyond the scope of this book, objectivity cannot be confined to the testing of the end-product of scientific research. This is not because the testing of the end-products is not important, for indeed it is of vital importance, but because the testing of the end-products embodies only one conception of objectivity and is only one component of the whole process of scientific inquiry. If anything, this phase of the discussion has been oriented around the premise that we require an intensive examination (i.e., a test) of the presuppositions underlying our current conceptions of tests and of objectivity.

The schema in effect suggests a combined sociological, psychological, and philosophical research program for the understanding of science. The schema is sociological in the sense that it explicitly embodies the idea of at least three different kinds of social inquiry roles — that of a philosopher P, a scientist S, and an object O. The schema thus allows one to bring sociological considerations to bear on the explication of objectivity. The schema is psychological in the sense that particular types will develop or have a preference for certain of the roles or modes of inquiry.[12] In other words, what role one assumes in an inquiry is as much a function of the choice and personality of the inquiring mind as it is a function of the problem situation. Thus the scientist, if he so wishes and is able, can choose to adopt the role of the philosopher, just as the philosopher if he so wishes and is inclined can choose to assume the role of the scientist in an inquiry. The assignment of roles is dictated as much by one's interests and functions in a particular

246

inquiry as it is dictated by formal training and professional interests.

The schema is also sociological and psychological in a number of other senses. For example, one of the most interesting questions raised by this study is: What are the detailed mechanisms whereby individual judgments or perceptions are combined into collective social judgments? As first proposed by Heider [178], the balance theory paradigm would seem to be a fruitful approach for future investigations to adopt in considering this question. Those who are familiar with the balance theory paradigm will perceive that it is a combination of the triads S_1 o_1 S_2 o_2 O and S_2 o_1 S_1 o_2 O; i.e., collective social judgments are a function of the interaction between S_1 o_1 S_2 o_2 O and S_2 o_1 S_1 o_2 O.

Finally, the schema is philosophical in that it explicitly incorporates the idea of different systems for engaging in inquiry and for resolving the judgments and observations of different observers. Thus, for example, even if the ultimate measure of objectivity were consensus, then, as Table 19 helps to demonstrate, starting with any IS as an initial state, there are many pathways that one could pursue in reaching consensus or a Lockean end-state. In sum, a general theory of science is dependent upon our ability to explore the consequences of the P o_1 S o_2 O in some systematic manner.

With the foregoing in mind, we can consider the proposition that strong emotions — the often hostile feelings between various types of individuals, the intense commitment to particular preferred positions — are not by themselves necessarily detrimental to the idea of scientific objectivity. It may well be, as the results of this book would strongly seem to indicate, that every scientist must be committed to his ideas if he is to be able to engage in particular types of observations, and that if emotional commitment blinds the scientist from seeing some things, then without it he would not be able to see other things. Some scientists are keen observers precisely because of their commitments, not in spite of them. The notion that bias and commitment are detrimental to the idea of science is not wrong — rather, it misses the mark. The mere fact that commitment and bias are present does not make an activity that contains them necessarily subjective. An activity is subjective if, and only if, there is no way of assessing the influence of some crucial factors on which that activity depends. Conversely, an activity is objective if we have a systematic way of assessing the key variables on which that activity depends. If, as the results

247

of this book strongly indicate, it is humanly impossible to eliminate all bias and commitment from science, then it is hopeless to pin our hopes for the existence of an objective science on the existence of passionless unbiased individuals. The task, then, is to show how objectivity is possible in terms of a conception of science that recognizes the active presence of commitment and bias, not in terms of a conception that dismisses them. This is the intent of the schema. It should not then be our goal to eliminate all emotion and bias from science, but rather to understand them, to assess them.

Under the conception of science developed in this book, to eliminate strong emotions and intense commitments may be to eliminate some of science's most vital sustaining forces. Perhaps every scientist has to be committed to his ideas so that they are developed and presented in the best possible light. And perhaps the system even needs a few individuals who are biased — who display the most extreme form of commitment so that they never shift, they never give up their pet positions. These few extreme individuals may be needed to insure that permanent advocates for every position, no matter how extreme or bizarre those positions may be considered to be at the time, survive to keep the idea of the position alive as a possibility. Too many ideas, which, when first proposed were rejected, have come back to haunt us as serious ideas. And in fact the best insurance against the proliferation of crackpot ideas may be the intense opposition of ideas by scientists of widely differing persuasions.

Clearly, though, if every scientist were committed to the same degree and kind as every other scientist, the idea of having objectivity depend on the process of one mind observing another would be absurd. If every individual shared the same commitments and biases as every other individual, it is difficult to see how one mind could ferret out the presuppositions of another. Indeed it is difficult to conceive how the notion of presuppositions would ever arise. The concepts *objective* and *bias* would be meaningless. Everything would be simultaneously objective/subjective, biased/-unbiased. The concepts objective/subjective and biased/unbiased only make sense because different observers possess different kinds and amounts of bias. Thus, although all knowledge may contain an element of bias, this is a far cry from the extreme position that all is bias. I would strongly object to that interpretation of the results.

The results of this study point to a conception of science that is

248

based on anything but the presumption of untainted, unbiased observers. Indeed, it contains a notion of scientific objectivity and scientific progress that is explicitly dependent on the presumption of intense commitments. The theme is a Kantian (IS) and Hegelian (IS) one: Science advances through the process of scientists of widely differing persuasions (types and degrees of commitment) thrusting their opposing conceptions and commitments at one another. Through this process science not only subjects its results to severe (but not crucial) tests but also exposes the underlying commitments of its practitioners.

It is important to emphasize that in this process commitments alone do not make for the objectivity of science. It is the presence of intense commitments coupled with experiments (Lockean IS), seemingly impersonal tests, arguments (Leibnizian IS), evidence, and general paradigms that make for the objectivity of science. No single one of the elements does the job. Science, as opposed to other systems of knowledge, is distinguished by the fact that, if not in theory then in actual practice, it has learned how to make use of strong determinants of rationality (testing, evidence, etc.) plus strong emotional commitments. And further, our characterizations and understanding of science have not kept pace with its actual practice. For this reason, I am opposed to relegating the study of science's emotional commitments to psychology and the tests of scientific knowledge to philosophy, because the study of the place and function of emotions and commitments in science is as much a philosophical problem as the nature of testing is a psychological problem. What is so disturbing about those such as Popper [347, 349] (see also Chapter One) who would relegate the study of one aspect of science to one field or another is that they take for granted the very distinction that is being challenged here — that the various aspects can be studied separately, let alone exist separately.

Likewise, I do not mean to say that science advances only through conflict, through the opposition of different types. If all scientists stood firm in their commitments, if no one ever shifted, general consensus would never arise. If science is marked by perpetual opposition in its ranks, it is also marked by periods of general consensus. Rather, the point seems to be that men shift their positions differentially and that the study of this differential shift is important if we are ever to understand how conflict gets converted into consensus. If we are honest, at this point in time we must confess we still do not comprehend the conversion process.

Science advances as much through the opposition of men of differing persuasions as it does through the opposition of different ideas. Science is more than just an abstract game of pure ideas. In this sense, men not only *learn* and *adopt*, in Kuhn's terms, the general intellectual paradigms of the age. They also learn (from their peers) and adopt specific sociological and psychological paradigms (styles, if you prefer) of doing science. These sociological and psychological paradigms are as much in opposition or revolution as the intellectual ones. Indeed, if every intellectual revolution is bound up with the personalities of its leaders, then it may make sense to ask not which comes first but rather how each aids and influences the other.

I can think of no more fitting epilogue to this chapter than the following passage from Churchman and Ackoff:

> One... point should be made with regard to the pragmatic meaning of "observation." The physical sciences have usurped a very narrow meaning of the term, i.e. they refer to what on a common sense level appears to be direct sensory experience. Such experiences are those of color, taste, certain smells, and the like. And positivism regarded these experiences as fundamental to scientific method. That such experiences are not direct in any meaningful sense seems to the pragmatist to have been well proved within modern psychology... But the much more important point is that *pragmatism does not consider "observation" to be restricted to such so-called elementary forms of experience.* A man's feeling at the given moment may have as strong a role in the process of reaching a conclusion as his sensation of sight. Pragmatism does not advocate a scientist who removes all his emotions, sympathies, and the like from his experimental process. This is like asking the scientist to give up being a whole man while he experiments. Perhaps a man's emotion will be the most powerful instrument he has at his disposal in reaching a conclusion. The main task, however, is to enlarge the scope of the scientific model so that we can begin to understand the role of the other types of experience in reaching decisions, and can see how they too can be checked and controlled. The moral, according to the pragmatist, should not be to exclude feeling from scientific method, but to include it in the sense of understanding it better [81, pp. 223–224].

Chapter Eight

Game of Science

RATIONAL, *adj.*: Devoid of all delusions save those of observation, experience, and reflection.
Ambrose Bierce in The Devil's Dictionary

"The rationality of natural science and other collective disciplines has nothing intrinsically to do with formal entailments and contradictions, inductive logic, or the probability calculus. Propositional systems and formal inferences are legitimate instruments, among others, for the purposes of rational investigation and scientific explanation, but they are no more than this. Instead, the nature of that rationality must be analyzed in quite independent terms, before the question can properly be raised, how 'logical systems' are given an application in science."
Stephen Toulmin [446, pp. 479–480]

A few years ago, in a series of two papers, the philosopher Bernard Suits raised some challenging questions: "What Is a Game?" [425] and "Is Life a Game We Are Playing?" [424]. The purpose of Suits's first paper was to lay out the general defining characteristics of all games. The purpose of his second paper was to argue that life could be viewed as a game, albeit a game that most of us played unconsciously and unintentionally. Viewing life as a game, Suits argued, allowed one to reveal some interesting things about the individual and collective intentions of the game participants that one could not reveal otherwise. At the very least, the possibility of viewing life as a game played by human participants allowed one to raise some rather interesting questions about human behavior.[1]

I distinctly recall my feelings toward the two papers at the time I first read them. I experienced dissatisfaction with the first paper because I felt that Suits had defined the notion of a game too narrowly. Even more, I remember being excited by the second paper. I was intrigued with the possibility of viewing science as a

251

game. Because one of my mentors, Thomas A. Cowan, had already seriously taken up this possibility [102], I was even more intrigued by the further possibility of combining and extending both Suits's and Cowan's ideas. Obviously, whether science literally is or is not a game is not the question, or at least not the question of concern here. It should be frankly acknowledged that the notion of viewing science as a game is a metaphor.[2] Hopefully, it is a fitting metaphor and one that is broad enough to allow us to capture and to sum up some of the distinctive features of science that this study has uncovered and emphasized. The contention is that the characteristics of a game provide us with a rich and broad enough framework within which to capture many of the more salient features of science.

The proper place to begin is with Suits's [424] definition: "I define game-playing as follows. To play a game is to engage in activity directed toward bringing about a certain state of affairs using only means permitted by certain rules, where the means permitted by the rules are more limited in scope than they would be in the absence of the rules and where the sole reason for accepting such limitation of means is to make possible such activity" [424, p. 209]. My principal objection to Suits's definition is that it is too narrow. More specifically, it focuses almost exclusively on the rules of games as their distinctive characteristic (as some philosophers and logicians of science focus almost exclusively on the logic of science as its distinctive if not sole defining characteristic).[3]

As a result, Suits not only downplays but indeed almost excludes all consideration of the role and behavior of the game players as a fundamental part of the game. I quote from Suits again: "Supreme dedication to a game [as in the case of a hypothetical golfer so devoted to the game that he seriously neglects his wife and six children] may be repugnant to nearly everyone's moral sense. That may be granted; indeed, insisted upon, since our loathing is excited by the very fact that it is a game which has usurped the place of ends we regard as so much more worthy of pursuit. Thus, *although such behavior may tell us a good deal about such players of games, I submit that it tells us nothing about the games they play*" [emphasis added] [425, p. 153]. I take strong exception to Suits on this point. I believe that the behavior of various kinds of game players tells us a great deal about the nature of the games they play. I also believe that the game players themselves form an inherent and essential part of every game. I do not believe that games are ever defined wholly or solely by their rules but

252

rather that they are at best partially contained in the complex interaction between the players and the rules.

I have further objections to Suits's definition. Consider that part of the definition which reads "and where the sole reason for accepting such limitation of means is to make possible such activity." Suits expands on what he means by this in the text of his paper: "In games I obey the rules just because such obedience is a necessary condition for my engaging in the activity such obedience makes possible. But in other activities — e.g., in moral actions — there is always another reason, what might be called an external reason, for conforming to the rule in question; for a moral teleologist, because its violation would vitiate some other end, for a deontologist because the rule is somehow binding in itself. In morals conformity to rules makes the action right, but in games it makes the action" [425, p. 154]. The above is not only naive psychology (as I intimated in Chapter One, this seems to be the occupational disease of philosophers), but it is also questionable epistemology. The motivations with which humans undertake any activity are too complex and manifold to ever be reduced to any single motivating reason. (Notice that if the definition of a game is fundamentally dependent upon the reasons, whether simple or complex, why game players accept the limitations imposed on them by the rules of the game, then the behavior of the game players — i.e., in the sense of their reasons — even in Suits's account does tell us something about the games they play.) From an epistemological standpoint, the question is whether any activity that involves humans can ever be defined purely independently of other activities. It is a fundamental characteristic of humans that they typically engage in one activity with some other activity or aim in mind. Indeed, humans never partake of completely isolated activities. Where one activity or game leaves off and another begins is thus a matter of no small concern or debate. In The Game of Science it is a matter of no small controversy whether The Game of Science is independent of other games — for example, The Game of Politics or the Pursuit of the Good.

For the purposes of this book, a broader conception of the notion of a game is required. This does not mean however that the concern is with the pursuit of a formal definition of a game or with distinguishing game-like from nongame-like activities or behavior. Instead, the focus is on noting the many aspects that enter into a broader conception of a game. I shall be primarily concerned with noting and commenting on the varieties of forms that these various aspects have assumed in previous accounts of The Game of Science. For my purposes then, a game is a special kind of human activity that is characterized by or possesses at least the following features, although not all these aspects are necessarily present in every game:

(1) individual players;
(2) coalitions, teams, sides, groups of players;
(3) the designation of one or more teams as the home side and the remaining teams as the opposing side or enemy team;
(4) a special field of play, a prescribed court, plus standard equipment (e.g., sanctioned league-approved balls and bats);
(5) entrance fees, entrance skills necessary before play can commence, ante, stakes;
(6) a schedule of games, home and away games, opponent vs. home court, special games, championship games, postseason tournaments, major and minor leagues;
(7) recruitment and development of talented players, farm clubs, training leagues;
(8) the rules of the game, the structure, what constitutes a move, a play, winning, losing, beginnings, endings;
(9) umpires, groundskeepers, scorekeepers, gatekeepers;
(10) aims, goals, objectives, ends:
(11) awards, prizes, special recognition, trophies;
(12) history and historians of the origin and development of the game, preservers of the tradition, establishment of a hall of fame, enshrinement, the designation and recording of great games, great players, game memorabilia;
(13) fans, general observing public, subjects, bookies, bettors, commentators, writers;
(14) commercial or public sponsors of the game, owners;
(15) the game as part of society, societal sanction or support of the game, legal, political, and social norms, laws governing betting, gambling, the sale and purchase of teams, players, playing courts.

It is relatively simple to list the preceding aspects of a game; it is much more difficult to discuss them in full.[4] Depending on the complexity of the particular game, the aspects of the game take on a multiplicity of different forms and meanings, and exist and interact at an increasing complexity of levels. For most games of any advanced degree of complexity, it is almost impossible to discuss

254

any one aspect independently of the others [16, 72]. Indeed, this is one of the major points that can be made about games. It is particularly true about one of the most complex of games that man has yet invented, The Game of Science. Unless one is a strict positivist, The Game of Science is not solely contained within the rules, the structure, or the logic of the game (aspect 8). Although the formal rules of a game are one of its most important aspects, they are not the whole of the game. No game is contained solely within any of its aspects. As a result, the explanation for any game cannot be given in terms of any one of its components. It must be sought in terms of the larger system of total game aspects. As a further result, no single discipline has a monopoly on The Game of Science (hereafter referred to as The Game). This means that no single discipline by itself is sufficient for explaining The Game.

The Game is intensely personal, intensely human. Stephen Toulmin wrote:

> Certain aspects of any science [its explanatory ideals, its specific concepts and theories, and the practical experience of its practitioners] can thus be described in impersonal terms: their historical development can be discussed as an episode in the history of ideas. Other aspects can be discussed in human terms alone: their historical development forms an episode in the history of human activities. At a more fundamental level, however, the intellectual activities of a scientific profession reflect both the stance with which the men concerned approach their experience of nature, and the concepts in terms of which they interpret that experience. At this level, we can no longer separate the activities of scientists sharply from the concepts and theories which are the outcome of those activities. In this respect, the central problems of an intellectual discipline are, at the same time, the central preoccupations of the corresponding profession. To reconstruct the historical evolution of "atomic physics" is, therefore, to trace the affiliations between the changing problems of successive decades and to show how the rational continuity of the subject was preserved throughout those changes [446, pp. 153—154].

Current written formal accounts of The Game (whether by scientists in scientific papers or by philosophers of science in treatises on The Game) bear as little relation to the actual conduct (plays) of The Game as high school civics texts bear to the actual workings of the political system. Personal elements are integral parts of The Game. As such they deserve to be incorporated into philosophical accounts of The Game to a far greater degree than they have previously been recognized and acknowledged.

I shall indicate here but two of the many ways that these per-

255

sonal aspects could be incorporated into contemporary philosophical accounts. The first way is severely critical of what has been variously referred to as "the orthodox view of scientific theories" [129, 180] or as "the received view of scientific theories" [426]. The second way is also critical, in this case of Kuhn's [247, 248] and of Feyerabend's [132—135] views. However, whereas the first way is severely critical and argues for the complete abandonment of the orthodox or received view of scientific theories, the second way merely argues for some strong modifications in Kuhn's and Feyerabend's views, not complete abandonment of them.

At some point every philosophical view of science is critically dependent upon the presumption of an observing scientist. More specifically, every philosophical view of science makes some assumptions about the nature of scientific observations or about observables and their relationship to a scientific observer. In the orthodox or received view of scientific theories, the assumptions take the form of assuming that there exists a class of entities distinguished by their capability of being directly observed by a human observer through his direct sensory apparatus. It is also assumed that there exist some theoretical entities which cannot be directly observed but which can be sharply differentiated from the directly observable of observational entities. It is further assumed that, although the theoretical entities are problematic because of their hypothetical and inferential status, the observational entities by virtue of their direct observational status are presumed to be nonproblematic.

In a previous paper [313] I have argued that the notion of direct observation or immediate apprehension is contradicted on every count by an overwhelming body of behavioral evidence. With respect to this evidence, the notion of direct observation is nonsensical. All the behavioral evidence I am aware of suggests that all observation by anything resembling a human observer is mediated by the observer's entire past behavioral and physical history, his current emotional state, his feelings, and his aspirations. In a word, all observations are observer dependent, dependent on the complicated and highly partial mental states of some observer. Observations are thus not neutral, because observers are not neutral with respect to what they observe. Indeed, for the act of observation to even occur, it is necessary that the observer bring with him some initial set of presuppositions about what he expects and even desires to see.

On two counts, then, observations are not theory-free, and so

256

they are anything but direct. First, in order to assess what any particular observer contributes to what he observes, we require a general behavioral and physical theory of observation which — relative to the class of all other observers — allows us to relate and compare the reports of different observers.[5] Second, in order to take any specific scientific observation we require some theory or model, no matter how crude or unarticulated (and not necessarily unique), of the event or process we wish to observe, particularly if our observations take place through complicated instrumentation. The results of this study more than support these assertions. Radically different observers or types of scientists bring radically different types of presuppositions with them to the field of scientific observation. Hopefully the results of this study also contribute to the eventual establishment of a behavioral theory of scientific observation that will allow us to compare the reports of different scientific observers. On a more abstract level, the results of this study support a presupposition-dependent view of scientific theories.[6]

As important as it is to raise severe challenges to the orthodox view of scientific theories, I believe that a more important task is to point out the modifications that the results of this study suggest to the views of Kuhn and of Feyerabend. Before these modifications can be presented, it is first necessary to recall briefly the essence of their respective views. Kuhn sees science marked by long periods of normal development (Normal Science) broken by occasional severe crises, or in his terms, "scientific revolutions." During a period of Normal Science scientists largely work within an accepted paradigm, an accepted, largely implicit world view. The accepted framework or paradigm provides relatively clear-cut guidelines for all the facets of scientific activity: it specifies the problems that will be worked on, the methods that will be used in working on them, the basic vocabulary that will be used for describing (recognizing) scientific problems, and above all a common language, process, or medium for resolving the inevitable disagreements or disputes that arise among scientists, data, and theories. Most scientists spend their life working within Normal Science, extending and working out the problems within the current paradigms. Very few scientists are either prepared to work on or actually do work within "extraordinary science" — that is, contribute to the overthrow of an accepted paradigm. For the most part, scientific training is training for Normal Science. There are times, however, when the current paradigms of Normal Science are chal-

257

lenged, threatened, and — even more rarely — overthrown completely by a new paradigm which in turn becomes the paradigm of the new period of Normal Science. Depending on how complete the revolution is, and apparently for Kuhn they are rather complete, the disjuncture between the old and new paradigms may become severe. Old and new paradigm proponents are unable to communicate across the gap; theoretical terms no longer possess the same meanings; they take on the meanings of the paradigms with which they are associated. As Toulmin has put it in characterizing Kuhn's views, "a scientific revolution involves a complete change of intellectual clothes" [446, p. 101].

Feyerabend emphasizes — as Kuhn does — the radical disjuncture between competing scientific theories. However, they argue it and reach it by very different pathways. Kuhn's pathways are historical and social psychological. He proposes radical disjuncture (or, as it has been called by some [236], "radical meaning variance theory") as a way to account for the fact of the growth and change in scientific theories, that science grows — contrary to popular myth — not by the patient and step by step accumulation of facts but by radical leaps and bounds. Feyerabend proposes radical disjuncture as a methodological principle, as a way of doing good philosophy of science: Scientists should always strive to produce the most radically differing counter-theories to the theories currently in existence. If all data are dependent on the use or presumption of some theory for their production and, as is often the case, the only theory we have available is the theory we wish to test, then the data produced through the use of that theory will constitute far less than the most powerful test of that theory. The test data will certainly not constitute an independent test of the theory [77]. Faced with this circumstance, the only way we can provide the strongest test of our theories is to test them against data uncovered by some counter-theory. The severest tests will be provided by the data derived from the theories that are in sharpest disagreement with one another.

Both authors have come in for some severe criticism. For example, both Kuhn and Feyerabend have been accused of overemphasizing the discontinuities, the disagreements between successive phases of the course of scientific development. They have been accused of placing a greater emphasis on the conflict between competing theories, paradigms, and scientists than actually exists or can be shown to exist. In short they have been accused of viewing the successive phases between competing theories or para-

258

digms as either perfect agreement or perfect disagreement. Dudley Shapere has written:

> It would appear that Kuhn's view that, in order for us to be able to speak of a "scientific tradition," there must be a single point of view held in common by all members of that tradition, has its source once again in the error of supposing that, unless there is absolute identity, there must be absolute difference. Where there is similarity, there must be identity, even though it may be hidden; otherwise, there would be only complete difference. If there are scientific traditions, they must have an identical element — a paradigm — which unifies that tradition. And since there are differences of formulation of the various laws, theories, rules, etc. making up that tradition, the paradigm which unifies them all must be inexpressible. Since what is visible exhibits differences, what unites those things must be invisible.
>
> Again, then, Kuhn has committed the mistake of thinking that there are only two alternatives: absolute identity or absolute difference. But the data at hand are the similarities and differences; and why should these not be enough to enable us to talk about more, and less, similar views and, for certain purposes, to classify sufficiently similar viewpoints together as, e.g., being in the same tradition? After all, disagreements, proliferation of competing alternatives, debate over fundamentals, both substantive and methodological, are all more or less present throughout the development of science; and there are always guiding elements which are more or less common, even among what are classified as different "traditions." By hardening the notion of a "scientific tradition" into a hidden unit, Kuhn is thus forced *by a purely conceptual point* to ignore many important differences between scientific activities classified as being of the same tradition, as well as important continuities between successive traditions. This is the same type of excess into which Feyerabend forced himself through his conception of "theory" and "meaning." Everything that is of positive value in the viewpoint of these writers, and much that is excluded by the logic of their errors, can be kept if we take account of these points [396, p. 71].

Kuhn in particular has been challenged to produce actual historical examples of radical paradigm switches, of actual cases where scientists were unable to communicate after a scientific revolution [396, 446].

Both Kuhn and Feyerabend have also been criticized on the grounds that their views lead to relativism — if each theory or paradigm defines its own terms, and if the disjuncture between strongly competing theories is complete, how can we compare theories? How can we say that one theory is better than another, and on what grounds are we justified in saying that progress has occurred just because one theory has replaced another? Further, if each theory supplies its own criteria, then, because of radical disjuncture again, how can we even compare criteria let alone theories. And even further, if the terms differ so radically from one theory to another, how can the data unearthed by one theory be used to test another theory? If data only have meaning with reference to their specific background theory, and if two theories differ radically in meaning, how can the data from one theory have meaning with respect to another theory and be used to test it?

To point out these difficulties is not to say that Kuhn and Feyerabend have not recognized these difficulties and acted to revise some of their views accordingly. Kuhn, in later accounts [248] has relaxed some of his arguments so that, for example, the disjuncture between rival paradigms is no longer as great; continuity between rival paradigms now becomes possible; communication between those on opposite sides of the revolution gap is now also possible. Further, a paradigm or a scientific theory is no longer conceived of as a tight, unitary system that must be either taken or rejected as a whole, but instead as a looser system in which partial changes are possible. In other words the replacement of one paradigm by another is not an all or nothing affair. However, granting this, the essential thrust of their arguments remains the same.

As much as I agree in principle with some of the arguments of the critics, I also agree on the basis of the results of this study. That is, my own data allow me to support a number of the critics' points with empirical evidence. And yet the critics are not entirely right either. There is a sense in which Kuhn and Feyerabend are right and a sense in which they are wrong or, better yet, incomplete. If, as their critics have pointed out, it is not an "all or nothing" choice between competing paradigms or theories, then it is also the case that Kuhn and Feyerabend are neither all right nor all wrong. There is a point in between.

The central issue *is* that of agreement. Kuhn is right when he emphasizes the tremendous conflicts and disagreements that exist among competing paradigms, theories, and scientists. The data of this study more than support him on this point. He is even right when he emphasizes that there are occasions for which there are theories, paradigms, and scientists that, for all practical purposes, may be regarded as in total disagreement. However, Kuhn is wrong when he either fails or neglects to point out that what is true on the micro level is not necessarily true on the macro level. That is, there are individual scientists who, for all practical purposes, may be regarded as in extreme disagreement with one another, and indeed the results of this study indicate that there are issues on which all scientists disagree some, and maybe even all, of the time. But there are also individual scientists who for all practical purposes may be regarded as in complete agreement with one another. Because both of these forces or tendencies operate in every system, it would seem false to characterize a system as existing in a state of total (or near total) disagreement or agreement.

There is a more telling way to put this point. It seems to me that the data of this study suggest that in every social system there are those kinds of individuals (types) who have a compulsive need to make revolutions, to disagree, as strongly as possible, with established ways of thinking — paradigms, if you prefer. These individuals have an almost consuming need to produce radical ideas and theories counter to those currently in existence if not in vogue. They seem to need to go out of their way to produce extremely novel ways of looking at old phenomena.[7] However the data also suggest that there are also those kinds of individuals who have a compulsive (security) need to preserve continuity with old established ways of thinking, to differ as little as possible from the tried and the true. Complete disagreement or agreement, rather than being actual states of affairs, are instead states of mind, attitudinal Ideals [16, 72, 408], or, in Toulmin's terms [446], the divergent "disciplinary aims" of radically distinct types of men. There are, in other words, scientists (Game Players) who act to produce, to bring about revolution in perpetuity, but there are also those who act to preserve the status quo, to bring about continuity in perpetuity. And there is an even larger body of players who are blends of these two extremes and more. Thus, to say that after a scientific revolution there is complete lack of communication between the two sides is to distort the situation. It is a partial truth at best in the sense that there are at any one point in time — before, during, and after a revolution — those scientists who are unable to communicate across any one of a hundred different gaps, theories, attitudes, issues. But there are also those kinds of individual scientists or forces acting within the system whose ideal is to discover links between the past and the present. It is too much then to say that after a revolution there is a complete breakdown between the two sides. This assumes a far greater degree of cohesiveness and homogeneity among sides, groups, or schools of thought than is warranted. I see as much diversity within the system or Game at any one point in time — even within supposedly tight common factions — as between successive time periods. As much as there are strong forces (revolutionaries) acting to produce drastic changes in The Game at any one point in time, there are also just as strong forces (reactionaries) acting to keep The Game the same as it has always been, and further yet, there are those (moderates)[8] acting to steer a course in between, to usher in the new while acting to preserve its links with the past.

There is an even stronger way to put these points and one that brings Feyerabend into the discussion. Michael Martin [276, 279] has raised a particularly interesting series of questions and challenges to Feyerabend's position. Although it is not exclusively associated with Feyerabend, Martin subsumes Feyerabend's position under the general heading of Extreme Theoretical Pluralism (ETP) which Martin characterizes as follows: "Develop as many alternative research programs [theories] as possible at all stages of development [of a science] whether the prevailing research program is progressive or not" [279, p. 5]. As a strong alternative to ETP, Martin lists another position which he labels Theoretical Monism (TM): "Do not develop alternative research programs in any science at any stage of development if the prevailing research program is progressive" [279, p. 5]. There are, as Martin notes, an infinite number of other positions in between. For example, for a particular set of (some) circumstances in the historical development of a science, one can develop some (not necessarily as many) alternative research programs whether the prevailing research program is progressive or not. As Martin notes, the interesting empirical question is: Which position is best for which science at which stage in its development?

A number of important implications follow from Martin's treatment of the issue. First, whether ETP, TM, or some other alternative actually holds or not is not merely a matter of empirical historical research. It would be extremely difficult for any finite set of observations on so complicated a phenomenon as science to establish a general rule like ETP or TM. Second, although rules like ETP or TM can be empirically researched, e.g., for the grounds of their empirical applicability, they are not solely or merely empirical propositions or hypotheses to be empirically confirmed or rejected. In a more fundamental sense, ETP and TM express the basic metaphysical or intellectual commitments of their proponents about the nature of science. They are not really empirical propositions about the nature of science. They are instead maxims or commands — directives — ("develop" or "do not develop") to view science in a very particular way prior to the gathering of empirical data on it. Rather than following from empirical obser-

262

vation and confirmation, they are instead prior to it. In this sense it is not a question of Feyerabend's being right or wrong but that his position merely expresses one stance out of the range of possible stances toward the nature of science (i.e., towards studying it, characterizing it, etc.).

One of the most interesting implications of the data of this study is that they suggest a novel way of studying various approaches to the nature of science. If one takes this study to its utmost limits, then one can assert that ETP, TM, and all the other alternatives in between are not just abstract intellectual possibilities but they are also psychological realities.[9] For every abstract philosophical position that can be formulated with respect to the nature of science, there exists a corresponding real scientist whose behavior and attitudes embody that abstract position. For every proposed strategy for playing The Game, there exists a corresponding type of Game Player who embodies that style of play. One way to study the implications of a proposed abstract position is to trace through its effects via a postulated, corresponding, human embodiment of that or any other position.

I am well aware that this position rests on a number of maxims or directives. That is, my own position is a particular approach to the nature of science. For example, one of the maxims that underlies my own position might well be labeled Extreme Interactionism (EI):

> Any proposition or assertion of any (every) field of inquiry has potential bearing for every other field of inquiry and in turn can be shown to rest on the results or disciplinary assumptions of some other field of inquiry. At every stage of the development of a field of inquiry one should always strive to make explicit and to develop these cross-impacts. The most radical tests of the concepts and propositions of any field occur when we attempt to apply them to fields traditionally defined to lie outside of their original domain.

I am the first to admit that there are an infinite number of other stances. They deserve exploration.

Kuhn was right in his basic appeal to social psychology and history as a way of doing philosophy of science. As Toulmin has put it:

> When Kuhn first wrote on this subject, American philosophy of science had been dominated for some forty years by a self-sufficient, anti-historical logical empiricism inherited from the Vienna Circle; and one real virtue in his work was to emphasize the need for a more historical and less formalistic approach to science. Evidently, the argu-

263

ments involved in changes of paradigm could not be described in the cut-and-dried terms of current inductive logic; one continuing strand in Kuhn's arguments has, accordingly, been an anti-formalistic one. A "revolutionary" change in the concepts of science is one too profound to be analyzed in terms drawn from formal logic alone [446, p. 126].

Kuhn went wrong, in my opinion, in his detailed application of social psychology. He was not social psychological enough to give a good accounting of the workings of scientists. For instance, Kuhn accords far too passive a role to individual scientists in bringing about a revolution [116].[10] As I have stressed, there are very specific kinds of scientists who actually seek to promote revolution.

Likewise it can be said that, for one who espouses radical or Extreme Theoretical Pluralism, Feyerabend was not radical enough in envisioning alternatives to his alternatives, how his ideas might be tested by other radical alternatives such as those outlined by Martin. However, these are minor complaints. Kuhn and Feyerabend still deserve much credit for opening up some needed new pathways in the philosophy of science.

Intense Masculinity of Game

So much has been said about this in previous chapters that little more needs to be said here. Perhaps all that needs to be done is to emphasize for one last time how deeply the masculine has permeated our entire conception of The Game. Not only the style of play but also some of our most fundamental root metaphors about the nature of The Game have been strongly masculine. This is certainly one way to interpret the popular conceptualization of science as a game against nature. It is even putting it mildly to say that the concept of nature as a hostile adversary against whom one must fight a constant battle and must eventually subdue (conquer) is the height of aggressiveness. However, whether it is masculine or feminine aggressiveness is almost beside the point as long as we finally recognize our metaphors for what they are — our own unconscious projective attitudes staring back at us through our creations. In another sense, though, it is important to recognize that it is fundamentally masculine aggressiveness, for then, finally, we may be able to ask what a science would look like that incorporated more of the feminine. Notice that I said "incorporated," for throughout this book I have not argued for *either* the masculine *or* the feminine exclusively but rather the blending and mutual support of the two. In support of this perhaps rather strange proposition, there is the evidence of this study plus the arguments by Holton [197, 198]. He suggests that some of the greatest Game Players in the history of The Game have had a strong feminine element in their thinking. The greatest adversary in The Game has not been nature but man and his concepts about himself and The Game.

Aims and Objectives of Game

These are as varied as the range of human aspirations. All too often philosophers of science have taken it for granted that the search for truth, explanation, or the increase in the empirical content of our theories — or some endless combination of these three — constitute the fundamental aims of science, and as a result they have been content to let it go at that. For example, in a recent paper entitled "A Critique of Popper's Views on Scientific Method" [284], Nicholas Maxwell argues as follows:

> My suggestion... is that we *should take* [emphasis added; notice the strong directive here] the search for explanations — for explanatory systematization — as a fundamental aim of science, and not as a derivative aim; not as an aim to be reduced to some more fundamental aim by means of some such claim as that explanatory theories are more verifiable, more falsifiable or better candidates for high verisimilitude than nonexplanatory theories. Explanatoriness should be conceived of as an end in itself, and not as a means to some other end.
> Given that the aim of science is to develop theories of both increasing explanatory power *and* increasing verisimilitude, how can we show that the methods of science give us the best, the most rational, hope of realizing this aim? [284, pp. 148—149]

I think that Maxwell is wrong in the sense that I believe that no end is ever an end in itself. I also think that Maxwell's characterization is incomplete. At best the aims Maxwell outlines are merely some of the many competing and conflicting aims of science. Surely The Game — indeed any game — does not ever exist in, of, and for itself; it always takes place within a social context, within a social and cultural environment; it needs support from that environment if it is to continue to exist. Surely, if to increase the explanatory power of our theories meant enslaving half the human race, the aim of increasing exploratory power could not continue

265

to be regarded as an end in itself, as worthy of pursuing in and of itself. Maxwell can only argue as he does because he takes too many other things for granted. As Martin has put it in commenting on Extreme Theoretical Pluralism (ETP):

> Now it should be noted that even if acting on ETP were to bring about maximal scientific progress it would not follow that ETP should be acted on. Suppose, for example, that a side effect of developing as many alternative research programs as possible in all branches of science were the loss of individual liberty or unacceptable state intervention. In short, suppose that scientific progress occurred at a maximum rate only when an undesirable political situation existed, e.g., a political regime which forced people to develop as many alternative research approaches as possible. This unacceptable political situation might outweigh any advantages that resulted from maximal scientific progress. Therefore, even if following ETP were to bring about maximal scientific progress this would not show that maximal scientific progress was on balance desirable [279, p. 5].

With regard to the existence of other aims, Russell L. Ackoff has argued that every social system is governed by — in some form or another — at least four aims or ideals:

> Men pursue *objectives* (ends or goals) by various *means* (courses of action) which incorporate a variety of conceptual and physical *instruments*. The capability for obtaining any objective presupposes the attainment of four ideals:
>
> (1) *The scientific ideal of perfect knowledge* (i.e., complete attainment of *truth*): the ability of every individual to develop and select instruments and courses of action which are perfectly efficient for the attainment of any end.
>
> (2) *The politico-economic ideal of plenty or abundance*: the availability to every individual of courses of action and the instruments necessary for them which are perfectly efficient for the attainment of any end.
>
> (3) *The ethical-moral ideal of goodness*: (a) the absence of contrary and contradictory objectives within each individual (i.e., peace of mind), since no state of plenty or knowledge can make the attainment of such objectives possible; (b) the absence of conflicting objectives among people (i.e., peace on earth, good will toward man), since only in the absence of such conflict can *every* individual attain his objectives.
>
> These three ideals would be vacuously attained if no one wanted anything. The universal wish for the ability to attain any objective itself presupposes that there are desired ends (i.e., unfulfilled objectives) and that they continuously expand. Hence, the last ideal:
>
> (4) *The aesthetic ideal of beauty*: the existence in every individual's environment of stimuli which inspire him to raise his aspirations, to enlarge the scope and meaning of his experience [16, pp. 430—431].

As Ackoff goes on to argue, these four aims or ideals are strongly coupled together — i.e., they can not be obtained independently of one another; furthermore, they have their counterpart within science itself:

> Consider the scientific ideal in isolation, the ideal of perfect and complete knowledge. There is clearly a politico-economic aspect of scientific progress. The resources required by science will have to be produced, distributed, and made acquirable, protected, maintained, and so on. There is also a scientific function involved in the pursuit of the scientific ideal: to provide science with the most efficient instruments possible in the development and use of knowledge. This science of science is precisely what *methodology* is all about.
>
> There is also an ethico-moral aspect to scientific progress. It involves the elimination of conflicts between scientific and nonscientific activities. One need only mention inter- and intradisciplinary conflicts or the conflict between science and religion to appreciate the need for this function. Finally the need for recreation and inspiration within science, the esthetic function, is also apparent. It is not accidental that a scientific work that inspires others is so often called beautiful. Nor is it accidental that there are so many scientific games (such as puzzles) to provide recreation to scientists.
>
> One could apply all four ideal-seeking functions to any of the four functions, not only to science as we have done above. Thus there is not only an ethical function in science, but there is a scientific function in ethics. If ethics has the function we have attributed to it and if the ULTIMATE GOOD is omnipotence for all, then science can clearly contribute to progress toward it and to measurement of that progress. *This does not reduce ethics to science* any more than science is reduced to ethics when we consider the dissolution of conflict relevant to its practice [16, pp. 245—246].

Thus, Ackoff believes, as I do, that science has at least four general or universal aims:

> The Knowledge Ideal of Science: *The scientific ideal of perfect knowledge...*
>
> The Politico-Economic Ideal of Science: *To provide every scientist with the physical and conceptual instruments necessary for the efficient pursuit of knowledge...*
>
> The Ethical-Moral Ideal of Science: *To remove conflict within scientists, among scientists, and between science and other social institutions...*
>
> The Esthetic Ideal of Science: *To enlarge the range and scope of questions and problems to which science can be applied* [16, pp. 430, 434, 437, 443].

The aims of science are unlimited, then, in the sense that they are as vast as the range of human aspirations, as varied as the objectives of diverse types of men. For example, if one takes

Jungian psychology seriously, then one can say that the aims of science as currently construed by many philosophers of science are too limited. It is not that the current aims are necessarily wrong but rather that they are incomplete. In Jungian terms, the current aims of many philosophers of science reflect the overwhelming dominance of thinking; the current aims are identified almost exclusively with the function of thinking (e.g., greater systematic explanation). However, if one takes Jung seriously, aims which arise out of each of the Jungian psychological functions can be conceptualized. [11] For example, Cowan argues, in his paper "The Game of Science" [102], for esthetic and intuitive aims of The Game. With respect to intuition, Cowan argues that "the course of scientific play leads to the full exercise of *intuition* within a prescribed framework of rules, the rules of scientific method" [102, p. 12]. Notice carefully again that it is not an either/or. It is intuition within the prescribed framework of thinking. I have not been arguing for the supremacy of one over the other but for their mutual incorporation and respect.

Concluding Remarks

To describe science as a game is not to demean it. [12] Even if science were literally a game, that by itself would not mean that it was not worth playing. The use of the notion of a game is intended primarily to call attention to the fact that science is a human creation; in the eyes of some The Game may be against nature, but science was created by man, not by nature. And if The Game is against anything it is probably ultimately against man.

Because The Game is intensely human and personal, and demands intense commitments, I do not mean to imply that The Game is entirely subjective, inherently irrational, or completely relativistic. The Game does have strong subjective elements; it does have strong irrational components; and it does have strong relativistic tendencies or aspects. But this is a far cry from saying that it is completely subjective, irrational, relativistic. [13] *As the author of this study, I would strongly resist any such interpretations of this work.* I have not done this study out of any desire to support those who lean toward an antiscience or irrationalist position. [14] I have few sympathies toward those who think the world would be a better place if we somehow got rid of all our science and technology. If Ackoff's aims are indeed fundamental, then I do not see how they can be accomplished without recourse to some form of science and technology. I do believe strongly that our science and our concepts of its structure, its aims, and its methods are in great need of revision. In this sense my motive has certainly not been to destroy or to hamper science, but to help improve it by means of an intensely critical study of it. Indeed, to the extent that I also consider myself a scientist, how could I think otherwise? In this sense at least I agree strongly with Popper: I identify one of the fundamental characteristics and aims of science with its critical spirit, its incessant need to study everything including itself in an unrelenting and critical way. What better way to study science than by science itself, one of the most critical forces that mankind has yet developed.

This study is not an attempt to demean the scientists who so graciously consented and cooperated throughout this study. From the beginning I was not out to disparage them or their beliefs. In many ways I both sympathize and empathize with their concerns. Even though I have been strongly critical of many of their beliefs, there is also much that I respect about those beliefs. And I certainly enjoyed the privilege of getting to know the scientists.

This leads me to a comment about the sample size of this study. Forty or so scientists is obviously a small sample on which to base such weighty generalizations about the nature of science. If as the cliche — and it is a cliche — goes, "90 percent of all scientists who have ever lived are alive today," [15] how can one found such strong generalizations on so small a sample? If the only legitimate basis for extrapolation or generalization is the size of one's sample, then the answer quite properly is that one cannot. However, I believe that there are other legitimate reasons for extrapolation. If for no other reason than to encourage and to challenge others to do similar interdisciplinary studies of science, I believe the results of this study should be generalized as far as possible. Philosophers of science do not hesitate to make all kinds of universal statements about the nature of science based on no empirical data at all, or worse yet, on the empirical data of what in their imagination they construe as the behavior of scientists. If philosophers are justified in doing this, on a statistical sample of zero scientists, then I do not see why one cannot extrapolate and generalize based on a sample of slightly better than forty. [16]

A word also needs to be said again about the potential charges of psychologism. This study is admittedly heavily psychological. However it is not in my mind to be labeled psychologistic because

I do not believe in psychologism as a philosophy. I have not tried to reduce every matter and concern of science to psychology. I do not believe that this can be done. Even more strongly, I do not believe this should be done. I do believe, however, that a personal and social element can be shown to be operating behind every aspect of science, that every aspect of science is heavily laden with exceedingly deep and personal elements. However, I also believe that a set of rational rules and public procedures also infuses the entire structure. In that sense, the intent of this study is not to replace the logic of science but to challenge it. I have tried to build the strongest case possible for psychology, not to make psychology the sole basis for studying a characterizing science. To replace the logic of science tradition entirely by the social psychology of science tradition would simply replace one reductionism with another. If, in spite of this, I have still been guilty of psychologism, then all I can say is that much of our philosophy of science has been guilty of logicism. If so, then this study is a needed counter to that overemphasis.

One of the main theses of this book has been that whatever the final picture of science, psychology has a fundamental role to play in constructing that picture. What are the psychological conditions that govern modern science? What are the types and qualities of mind that give themselves over to its spirit? Not everyone prizes the qualities of mind, or the social conditions and institutions, that support such minds, that go into the making of a scientist. It is important to know why. It is important to know about the nature of those that do.

Above all I have tried to argue that we can no longer afford to ignore psychology in any future accounts of science. We do so at our own peril, at the peril of insuring our continued ignorance about ourselves. What are the forces that drove man to create science? When will we realize that one of the greatest purposes of The Game is as much — perhaps even more so — to learn about ourselves as it has been to learn about nature. This should have been one of the major purposes behind the planning of the lunar missions.

Long ago Immanuel Kant wrote at the conclusion of his *Critique of Practical Reason*:

> Two things fill the mind with ever new and increasing admiration and awe, the oftener and more steadily we reflect on them: the starry heavens above me and the moral law within me. I do not merely conjecture them and seek them as though obscured in darkness or in the

270

transcendent region beyond my horizon: I see them before me, and I associate them directly with the consciousness of my own existence. The former begins at the place I occupy in the external world of sense, and it broadens the connection in which I stand into an unbounded magnitude of worlds beyond worlds and systems of systems and into the limitless times of their periodic motion, their beginning and their continuance. The latter begins at my invisible self, my personality, and exhibits me in a world which has true infinity but which is comprehensible only to the understanding — a world with which I recognize myself as existing in a universal and necessary (and not only, as in the first case, contingent) connection, and thereby also in connection with all those visible worlds. The former view of a countless multitude of worlds annihilates, as it were, my importance as an animal creature, which must give back to the planet (a mere speck in the universe) the matter from which it came, the matter which is for a little time provided with vital force, we know not how. The latter, on the contrary, infinitely raises my worth as that of an intelligence by my personality, in which the moral law reveals a life independent of all animality and even of the whole world of sense — at least so far as it may be inferred from the purposive destination assigned to my existence by this law, a destination which is not restricted to the conditions and limits of this life but reaches into the infinite [217, p. 162].

We have developed the kind of science (Apollonian) that knows how to reach "the starry heavens above." We have yet to learn how to develop the kind of science (Dionysian) that knows how to reach "the moral law within." We have yet to appreciate fully that The Game is nothing if it is not supremely esthetic and moral [102].[17] This is the task ahead. If we fail, as we have in the past, then, in the words of Arthur Koestler [234], we shall remain, like the founding fathers of modern astronomy, "intellectual giants but moral dwarfs."

Notes

Chapter 1

1 See Norman Mailer's *Of A Fire On The Moon* [270, p. 471] for the reference to "marvelous little moon rocks."

2 Although the social critics [99, 321—324] and the proponents of science may differ over whether the landing of men on the moon represents the sustained advance of progress, or instead the cult of progress in its highest form, they seem to agree that it is one of the direct by-products of the sustained growth of science.

3 For those readers who might feel that I am being unduly pejorative or overly emotional in my use of the term *Storybook*, I would point out that no less a distinguished sociologist of science than Robert Merton has seen fit to use the same term in much of the same spirit and connotation implied here [297, p. 16]. As strong as the term Storybook is, it is not strong enough in my opinion to counter this naive view of science.

4 For an extended argument on precisely this point, the reader is referred to the delightful book by the British physicist A.M. Taylor [433, pp. 3—4].

5 It might be comparatively easy to dismiss the Storybook version of science as an outrageous caricature if the accounts of supposedly more sophisticated observers did not have a disturbingly similar quality. For an example of what I mean, I refer the reader to a book by the eminent historian of science George Sarton, *The History of Science and The New Humanism* [387, especially p. 53 and pp. 106—107].

6 The contention of this book is certainly not that Robert Merton in particular is oblivious of the disparity between the actual behavior of scientists and the idealized image of them as portrayed in the norms. Nor do I contend that Merton has not revised his thinking with regard to the norms [291, 292]. The contention is instead that Merton has not used this discrepancy to make substantial revisions in his initial formulation of the norms [297]. Nor does it seem to me that he has adequately reflected on how these very norms might themselves have hampered study into their violation. Appearing as they do as the very epitome of scientific rationality, it may have been exceedingly difficult for sociologists to have imagined alternative normative structures for science, let alone for them to have done research directed toward the explication of these alternative structures.

7 See, for example, Storer [423, pp. 82—83].

8 For example, the point has been made [421] that the current norms are as much a manifestation of the ideology of their proponents as they are of the actual state of the social system of science. Maurice R. Stein, for

one, has argued that there runs throughout all of Merton's work a preference for a certain style for the doing of sociology. In particular, Stein accuses Merton of "anti-poetic militancy" and of defeating his own arguments for the use of neutral, unpoetic language in formulating sociological theories "by the partisan vigor with which he [Merton] repudiates opposing conceptions" [421, p. 177]. Those with opposing conceptions of sociology and of science may very well formulate different versions of the norms. In this sense the norms of science thereby become an interesting problem for the sociology of knowledge.

9 For a systematic statement of this point of view, see Popper's *The Logic of Scientific Discovery* [347, p. 31].

10 See Popper [348, pp. 57—58] for a statement as to the superiority of the logic of science in comparison to the social psychology of science.

11 My point is that every logic of science has embedded within it an implicit psychology and sociology of science, and vice versa. The choice therefore is not between the logic or the psychology or the sociology of science, as Popper [348], for one, has posed it. Rather the choice is between a logic, a psychology, and a sociology of science that incorporate the best, most informed features of one another and those that incorporate the worst. My contention is that unless a more sophisticated sharing and incorporation of knowledge takes place, all of these fields promise to remain as little more than extreme specializations, studying not science as it actually is, but rather, their own narrow, self-constructed caricatures.

Consider the above within the context of the issue of psychologism. If by psychologism one means that psychology is perceived as the basic science and that everything can fundamentally be reduced to psychology, then I quite agree with the charge of Popper and others [318] that this conception of psychology is psychologism and is a dubious if not dangerous philosophical doctrine. The serious problems of philosophy are no more reducible to mere matters of psychological fact than are the serious problems of any field. But if the charge of psychologism is meant to refute any application of psychology or to deny that every situation can be scrutinized for its psychological components (as well as for its logical, esthetic components), then I must strongly take issue. If psychologism and sociologism are the occupational diseases that psychologists and sociologists, respectively, are prone to, then the strength with which logicians oppose almost any application of psychology (or with which they demean it by relegating it to second place behind logic) makes them liable to the charge of logicism [446].

Perhaps the best way to meet the charge of psychologism is to show that, try as one might, one can never avoid all references to psychology and that therefore those who oppose psychologism cannot really be against all psychologizing but only a psychologizing of a particular kind. I refer the reader to a previous essay [309] where I attempt to show this.

12 See footnote 10.

13 I quote from Churchman: "The project [a Churchmanian program in the philosophy of science] can be characterized as follows: to study the procedures of science without assuming any hierarchy of the sciences. This means that no discipline is to be regarded as fundamental with respect to another. Or, in specific instances, there is nothing in the nature

274

of one problem as such which required that it be studied prior to certain other problems" [71, p. 257].

14 See Popper [347, p. 31] for a prime example of this kind of insulating tactic.

Chapter 2

1 The categories NASA and USGS are not lumped together, even though they are both governmental agencies and even though they perform work that overlaps at many points. Nevertheless, they have distinct institutional histories and characters, a point which repeatedly comes out in the interviews. For this reason, they are reported as separate categories. On the other hand, governmental agencies such as the Canadian Geological Survey, which is very similar in character to the USGS, would be included under the USGS category. Likewise, governmental research laboratories — like the Brookhaven National Laboratory — that have an identity which is distinct from that of the USGS or NASA would be given a category of their own. The Canadian Geological Survey and Brookhaven Laboratory are used only as examples; no scientists affiliated with these institutions were interviewed.

2 Although it is difficult to get a precise determination of the exact numbers of PIs and Co-Is in the total population, one can form a rough estimate. From various technical magazines and internal NASA documents one can set the total number of PIs and Co-Is at approximately 314. Of this number, approximately 62.8%, or 197, were university-based PIs and Co-Is; 10.2%, or 32, were NASA affiliated; 13.7%, or 43, were USGS affiliated; 6.0%, or 19, were affiliated with government research laboratories; and finally, 7.3%, or 23, were from industry.

3 You can perhaps appreciate this point better if you realize that it costs dearly in both time and money to ask questions. The facts of social science, like the facts of all science, are not cost-free. In the case of a project like the Apollo missions, where it cost literally billions of dollars to collect the basic scientific facts, this should be patently obvious. As Levinson and Taylor [261] have so graphically put it, "Approximately 24 billion dollars had been spent on the Apollo program before the first rock was returned ($500,000,000 per pound)" [261, p. ix].

In this study, I had only so much time with each respondent to ask but a few of the very many questions that I would have liked to have asked. And like every researcher, I had only so much money to spend in accomplishing my purpose. Every question had to be chosen carefully to minimize expenditure of both time and money, but hopefully to maximize information gained. Hence I had to get as much out of each question and each respondent as I could.

Chapter 3

1 See the bibliography at the end of this book for a selected listing of some of the more technical papers in the area.

2 *See also* [281, 404].

3 The reader is referred to the following sources for a more extended discussion on the differences between well-structured and ill-structured or elusive problems [282, 306, 310, 314].

4 The notion of *plausibility* was not used in any technical sense. It was not defined for the respondents in any precise way prior to their rankings. As a result, its meaning was most frequently taken as equivalent to that of *probable*. For a more precise notion of plausibility that does differ from that of probability, see Nicholas Rescher [361]. Rescher opens up the possibility of doing future studies which make it their goal to measure the relative plausibility of a theory as distinct from its probability. He demonstrates that the calculus for plausibilities is not the same as the calculus for probabilities.

5 This comment is interesting because it corresponds almost exactly with some of Murray Davis' notions [116] of what makes a theory or point of view interesting. For a theory to be interesting, it has to be different enough but not too different or its proponent will "get labeled as a crackpot." See [116] for details.

6 It is important to point out that heavy emotions were not just concentrated on one side. Although in the extreme minority, those who saw little that is positive in commitment and bias could be just as strong in their feelings regarding science as those who saw something positive in commitment and bias: Scientist X: "There is a certain virtue in the interplay of ideas in which some people are the champions of a particular point of view... Those on the other side put forth all possible arguments against the theory on the one side. However, in general it is dangerous for people to have strong feelings whether or not they are objective." In general, X was rather strong in his feelings that it was "dangerous for people to have strong feelings."

7 Commitment in this sense is clearly different from the kind that Merton considered in his original formulation of the norms of science [297]. This may help to explain why Merton [297], unlike Barber [33] and West [468], did not take emotional neutrality as a norm of science. Merton [297] is right in saying that science demands commitment, commitment to the ethos of science, and in this sense the notion of the emotionally uncommitted scientist is absurd. However, it is important to point out that in later accounts Merton [291, 294] considers commitment in the sense that it has been considered here — that is, commitment to one's hypotheses as a necessary ingredient in the doing of science. In the sense considered here, emotional neutrality toward one's scientific ideas would be the norm of science; emotional commitment to one's scientific ideas would be the opposing counter-norm of science.

8 However, if a recent article in *Science* on the Piccioni lawsuit is any indication, stealing, or at least the belief in stealing, may not be as rare as we may have imagined: "Various sorts of cheating among scientists have been alleged since Newton's day, and the suspicion of cheating has been a frequent subject of study by historians and sociologists of science. A survey [143] in 1967 and 1968 of over 200 British high energy physicists, for example, found that over one-sixth of them believed earnestly that some of their work had been stolen at some point" [399, p. 1406].

276

If this is true, then we need better ways of dealing with this unpleasant phenomenon than pretending it does not exist or by sweeping it under the rug. *See also* footnote 15 for another scientist's view of this matter. Roughly a fifth of the sample referred to stealing as a problem in one way or another. The postulation of a norm like secrecy is thus not based on the isolated experiences of a single scientist.

9 Scientist Smith said: "Zimmer deliberately prevented Park from getting support to do research on phenomenon Q. This is suppression. The case of discovery Z was stolen from Park without any acknowledgment. It's hard to say whether this was deliberate."

10 Hints of this norm were reflected in the comments of the scientists. One of the scientists (Rommer) who had been in a special position to know elaborated in great detail on the political machinations (from his point of view) that went into the selection of some of the PIs. According to Rommer, some of the PIs got some extra lunar material over and above that actually required for their basic analyses. As Rommer put it, "It's probably just as well that the best guys who know how to use the material got it." Said Rommer: "Sometimes a guy known by no one submitted a proposal. Usually then the proposal mattered. *In most cases the proposal didn't mean a damn thing* [emphasis added]... In general, it was a pretty fair process. Some things were simply influenced by the membership of the selection committee, but it wasn't a matter of a conscious biasing... There were obvious examples of bias or politics but that wasn't the rule of the day.

"A lot of interest in the samples was due to the fact that NASA was funding large amounts of money. *A lot of people still involved in the program aren't really interested in the samples* [emphasis added]. I don't think it's prestige or status since most of these guys are already well known. Their reputation isn't going to be affected by whether or not they analyze a lunar sample. The reason is funding. I'm really bothered about Scientist Bracken in this regard. Bracken has really milked NASA for an incredible amount of money. But then NASA probably would have wasted this money anyway so I'm just as glad to see it go to Bracken where at least something good for the profession is likely to come out.

"I suspect that a lot of samples were distributed on the basis of an unspoken agreement because I'm sure guys like Jones and Smith wouldn't have participated if it had been blatant. But they [Smith and Jones] really took care of themselves. Quay ended up with large amounts of q type rocks and he hadn't even proposed specifically to work with q type materials. There was a great deal of the distribution of samples to one's fellows. *The amount of material one got wasn't just solely related to the amount one needed* [emphasis added]. There was an effort made to give a large chunk to a guy who needed one. I don't think anyone was issued a great excess of sample, but some got a lot more samples, which enabled them to synthesize more data and get a bigger picture than others. Adams and Baker got a lot of rocks. There were some inequities in the distribution but I'm not too unhappy with it as those guys had a nasty job and worked hard and are entitled to a little extra in the way of samples. There are guys who are very bitter about it, for example, Nolan. Oaks probably feels he got a lot of samples but that he didn't get enough [this is repeated].

277

Oaks bent people's arms to get more as he and others were finding terribly interesting things but then he presented a miserable paper [at the Houston Apollo 11 conference]. Hall is probably upset about the whole business. Hall has flashes of brilliance occasionally and flares up but usually he is rather a clod."

11 The norm of partiality is a direct expression of the Churchmanian idea that the moral consequences of science are inseparable from its technical features [70, 72, 77, 79, 81]. In Churchmanian terms, it is pragmatically infeasible to distinguish between the two. It is not that the sociologists' norm of impartiality is all wrong and that the Churchmanian norm of partiality is all right, but that the sociologists have seen no need for a more reflective justification of the norm of impartiality. What apparently for the sociologists of science is a natural fact of scientific life is for me only a postulate that serves to indicate the sociologists' conception of science. Similar points could be made with respect to each of the other norms.

12 Strictly speaking, it does not make sense to refer to the *geology of the moon*. The correct term is *selenology*, which derives from the word *Selene*, the goddess of the moon in greek mythology. However, because the expression geology of the moon is more widespread than the more esoteric term selenology, I shall continue to refer to the geology of the moon. I trust that this note will help to eliminate any possible confusion in meaning.

13 The field of geology provides an excellent setting for so many of the topics with which this book is concerned. For example, the proceedings of the 1928 meeting of the American Association of Petroleum Geologists, *Theory of Continental Drift* [8], is a virtual sourcebook for many of the attitudes that are being discussed. The problem of continental drift has been as ill-structured and controversial a problem as the origin of the moon. And so, like all important problems, it has brought forth a number of archetypal attitudes toward the handling of the problem. As was to be expected, each of these attitudes toward the treatment of the problem was itself indicative of a more fundamental underlying attitude toward the nature of science.

A few of the respondents in the sample were earlier involved in the continental drift controversy and were thus in a position to testify on the differences, in their opinion, between those who quickly accepted the hypothesis of continental drift and those who resisted to the very end. The difference seemed to be that those who were quick to adopt the hypothesis were mainly impressed by global and intuitive evidence that was accumulating from a variety of sources. Those who resisted required a precise and detailed verification of the exact mechanisms and processes by which the continents could drift before they could accept the hypothesis. The actual situation was, of course, much more complicated than this simple difference between two types. As one of the respondents suggested, some investigators just happened to be in the "right place at the right time," so that they had access to the crucial evidence earlier than their colleagues. However much this was this case, there still remains the fact that the continental drift situation provides a vivid illustration of the differences in attitude we have been exploring. As one leafs through the pages of the 1928 proceedings, one cannot help but be struck by the fact that we are dealing with very different conceptions regarding the nature of evidence and of science.

For an eminently readable discussion of the continental drift controversy in terms the layman can readily comprehend, I recommend *Debate About the Earth: Approach to Geophysics Through Analysis of Continental Drift* [431].

14 The approximate status of a scientist was computed by weighting such characteristics as the number of significant awards he had won, the number of important professional offices held, the prestige ranking of the institution where the scientist was then located, the prestige of the institution from which he had received his highest degree, the number of times he was cited in the *Science Citation Index* [6], and the number of times he was referred to by his peers in the sample. Although such a procedure obviously does not give a perfect or an absolute measure of status, it does allow one to form a rough estimate of the various status groupings of the scientists. In addition to these factors, there were more than enough side comments from the respondents to infer their status rankings of the sample. And in fact, the rankings of the respondents were given the highest weighting of any factor in computing the rankings. This was done on the presumption that for so judgmental a variable as status, they were probably the best judges. As a result, the status rankings are best described as "perceived or judgmental rankings." They certainly are not absolute in any sense. Thus, it should come as no surprise that the perceived type of a scientist and his perceived status are highly correlated. The high correlation says that the two attributes of status and type are perceived attributes which for any respondent are highly associated.

15 The distinctive characteristic of this group is that they experienced shock. It is not when they experienced it. The following quote from one of the scientists (Quay) is interesting not only in this regard but also for the insight it gives about one man's motives for becoming a scientist: "I thought professors were gods through my first year at the university. When I found out they were just ordinary people it shook me up. I don't know how I coped with it. My faith in science was shaken. Becoming a scientist was like my decision to be an atheist after having been raised in a very religious family and not believing in it [religion] from age two... Becoming a scientist was to me a missionary thing, not a matter of earning a living. I was an X kind of scientist when the name wasn't even known. I wanted to become an astronomer, and even wanted to work on the moon. I wanted to do a thesis on moon at Z but I wasn't permitted to because of the tunnel vision of others; it wasn't regarded as an appropriate subject for geologic contemplation. You couldn't do field work. Work would have to be based on photos from 340,000 mi. [When I asked if science had for Quay what religion did not, and if this was why the shock that professors were not gods was so traumatic, Quay said again but in a very indirect way that he wanted to do something with a missionary zeal.] In those days it was ridiculous to want to become an astronomer; it's the same as wanting to be a poet today.

"There is too much conservatism now in science. It is built into the education and into the system, in formal organizations, particularly governmental ones. It I had worked with USGS, they would have screened out a considerable number of the papers I've written, because they are so conservative. Their screening processes seem like small things, but they are really tremendous fences. [When asked what made scientists conservative, Quay said that he was not sure.] It's all relative. Science is more conservative than the layman would suppose, than the standard account talks about. People in science are much more concerned with petty jealousies and security than most people would think. And it's even more true nowadays. People are in science as a livelihood, and largely are not missionary spirited. Once you get Big Science, you let in the rank and file. The causes of the conservatism of science are that scientists are generally specialized, and are not well read, or trained in the humanities. Very few scientists read anything outside of their science after college or even in the other sciences. The pressure is to know more and more about less and less. It's more difficult now to attempt to become a spectrum scientist. The easy way to be successful is to be a specialist, when you can be the best in a field of work, and you have only a bit of literature to read.

"I think your study is extremely vital. It's important to dispel some of the misconceptions people have about the scientific method and how science works. It is also important to promote the social sciences. Scientists aren't as misinformed about themselves as the general public is.

"They have been sitting on a paper of mine for four and a half months at X [an important journal]. I will present the paper tomorrow, but I won't present two or three important aspects of the paper. I just don't want to give away the method whereby I've obtained the knowledge to where Z's were. There are certain more or less piratory [sic] things whereby I've obtained this knowledge. [When I asked if this was in conflict with the scientific norm of the open sharing of ideas and data, Quay said no.] There just is no freedom of sharing ideas. Ford told me he knows where a new P structure is, but he won't tell me where, and I haven't pressed him, but I told him to publish it. Ford should publish it before someone else does. I don't like to give results out unless they are coming out within the next 6 months.

"I still have a missionary zeal for science. I still think it is the solution to all our ills. The scientist can contribute most by doing his job as a scientist. I can see why people in the humanities might want to become activists, because they don't know where they are going. I know where I'm going, what's important to do.

"If Newton came back to earth, he would be surprised at what science has accomplished. But if Rembrandt or da Vinci came back to the Metropolitan Museum, what would they think? In spite of the imperfection of science we know we are building a pyramid, laying brick on brick. The worst thing you can do is to write something which is wrong. The worst thing that can happen to you when you write a paper is for your paper to be completely ignored. If a person objects to it, that's fine. If they agree with you it's even better."

280

Chapter 4

1 In a section entitled "Objectivity," Osgood and his colleagues write: "A method is objective to the extent that the operations of measurement and means of arriving at conclusions can be made explicit and hence reproducible. The procedures of measurement with the semantic differential are explicit and can be replicated. The means of arriving at results, from the collection of check marks on scales to the location of concept-points in semantic space and the production of conceptual structures, are completely objective — two investigators given the same collection of check marks and following the rules must end up with the same meanings of concepts and patterns of conceptual structures. It is true that how one *interprets* these results is a subjective matter, but so is the engineer's interpretation of objective data on the stress which a bridge will stand. It may be argued that the data with which we deal in semantic measurement are essentially subjective — introspections about meanings on the part of subjects — and that all we have done is to objectify expressions of these subjective states. This is entirely true, but it is not a criticism of the method. Objectivity concerns the role of the observer, not the observed. Our procedures completely eliminate the idiosyncrasies of the investigator in arriving at the final index of meaning, and this is the essence of objectivity" [338, pp. 125—126].

It is a shame that Osgood and his colleagues think that the "essence of objectivity" consists in "completely eliminating the idiosyncrasies of the investigator in arriving at the final index of meaning." I argue that we should not necessarily wish to eliminate the idiosyncrasies of the investigator, which is an impossibility anyway, but rather to control for them by studying them. The thing that is so disturbing about the Osgood concept of objectivity is that it prevents us from studying those idiosyncrasies and hence from learning about alternate notions of objectivity which do not depend on the "complete elimination of the idiosyncrasies of the observer." Osgood wants to eliminate the very thing that is of interest to the social psychologist of science and which may be of prime importance to science.

2 It might be contended that this has already been done [418]. However, a detailed inspection of the literature supports the contention that such an investigation or exercise has not been done in the sense reported on later in this chapter. Even when subject protocols have been collected, they have been interpreted from the narrowest of paradigmatic bases; for example, subject protocols have been interpreted in terms of classical conditioning theory. To analyze the protocols in terms of classical conditioning would not only seem to rob them of their power but to take away from their status as meaningful information in their own right which deserves to be taken on its own level. A more relevant exercise for future analyses would be the analysis of verbal protocols by content analysis. Content analysis would at least attempt to deal with the conceptual information that was present in the protocols. It would certainly not demean it by lowering it to the level of classical conditioning responses.

3 For reasons of space consideration, the various questionnaires that were administered are not reprinted in this book. These instruments are available from the author for inspection for those interested.

4 The results and conclusions of this section are borne out by formal factor analyses of the quantitative data. For reasons of space, these analyses have not been presented here.

5 See Walter Weyrauch's fascinating study, *The Personality of Lawyers* [471]. There is an uncanny similarity between many of the attitudes and personalities of Weyrauch's sample of lawyers and the sample of scientists studied here.

6 From the means of the sample responses on each of the 10 dimensions in Figure 3, one can compute a general Euclidean distance between any two of the Individual Scientists [337, p. 90]. From a matrix of distances or proximity measures so computed, one can then compute the minimum number of dimensions that "adequately reproduce" the rank order of the original distances between the Individual Scientists [227].

7 For example, the responses to the items dealing with the scientists' religious backgrounds and preferences indicate that the majority come from Protestant backgrounds but either have no religious affiliation themselves or hardly ever attend religious services, although a considerable percentage of them feel the need to believe in something.

8 It has been suggested to me on numerous occasions that the use of better adjective pairs (opposing pairs) would have produced a preference for one end of the scales, contrary to the situation that resulted. Although this is true, it misses the point entirely. It is just as important to find out what scientists do not consider as opposites as it is to find out what they do. Besides, before the fact and even with a pre-test, I had no sure way of knowing that such a significant body of the sample would check both ends of the Cognitive Style scales. Surely the fact that so many scientists felt the need to check both ends of these particular scales is significant in itself. It is always possible to get differences by making finer and finer distinctions. But is this necessarily what one wants? I find it especially disconcerting that there were so many social scientists who are unable to appreciate this point. The search for differences is not necessarily to be prized over the search for similarities.

Chapter 5

1 The statements in Table 1 are exceedingly general and nonspecific. For example, crucial terms like *hot* and *considerable* are undefined. This was done purposefully, to draw out the respondents' meanings of the terms. One of the things I was interested in was precisely how they would react to open-ended attitudinal statements about the moon. Many of the respondents felt it necessary to specify or to qualify a number of the statements before they could respond to them.

2 In a way this was done. Each respondent was given the opportunity to write in other issues that were of particular interest to him. The responses to these issues are not reported here because they are more easily and systematically dealt with in the next chapter.

3 The analysis here is only approximate, because direct probability judgments were not collected from the respondents. Instead, the Likert re-

sponses of Table 1 were transformed into the probability judgments of Table 2 via the straight-line transformation "1" = probability of 1, "4" = probability of 0.5, and "7" = probability of 0.

4 By this or any analysis, I do not mean to imply that the Apollo missions produced no significant alteration at all in our scientific beliefs. I certainly do not mean to imply that the Apollo missions produced no new or significant scientific information about the moon. Nothing could be further from the actual case in point. If anything, the missions have produced too much information, so much that the respondents continually remarked that just to assimilate the information already produced would keep them occupied for years after the Apollo missions had ended. The point becomes even stronger as soon as one realizes that not all possible information has been gotten out of the returned Apollo samples. Only a fraction of the total samples have been analyzed [146].

What I do mean to imply is that, in terms of the analyses in the text, the information transferred is small or large depending on how one looks at it, that is, on how one conceptualizes the phenomenon. The purpose of the analyses in the text is to raise questions and challenges, not to state flatly that the information transferred by the Apollo missions was small. Obviously the information transferred depends on a host of variables, for example, the particular issues sampled. These variables deserve further and systematic exploration. This is the purpose of the analysis, to confront this issue.

5 This statement deserves an important qualification. Analyses of covariance failed to pick up any significant differences between the positions of the different types of scientists on any of the 11 issues of Table 1. However, one-way analyses of variance picked up three cases of significant differences out of a total of 33 possible cases (from Table 1 there are 11 issues across 3 time periods and thus 33 cases). Two of the cases are significant at the 0.001 level, a high level of significance; the third is significant at the 0.05 level.

6 Consider the following statement by Jung: "In medical circles the opinion has got about that my method of treatment consists in fitting patients into this system and giving them corresponding advice. This regrettable misunderstanding completely ignores the fact that this kind of classification is nothing but a childish parlor game, every bit as futile as the division of mankind into brachycephalics and dolichocephalics. My typology is far rather a critical apparatus serving to sort out and organize the welter of empirical material, but not in any sense to stick labels on people at first sight. It is not a physiognomy and not an anthropological system, but a critical psychology dealing with the organization and delimitation of psychic processes that can be shown to be typical" [215, pp. xiv—xv].

7 Those who are familiar with the Jungian system will note that I have neglected Jung's important distinction between introversion and extroversion. I have done so for ease of exposition and of measurement. That is, if one regards introversion and extroversion as independent dimensions, then there arise 8 psychological types, not 4. Thus, one can be an introverted or an extroverted thinking, feeling, sensation or intuition type. These further distinctions are neglected here. It would have required too

much time to have the respondents read and rate 8 portraits instead of 4.

8 For example, consider the following: Scientist A: "These are descriptions of tendencies. Every person I can think of would not fall neatly into these categories. The categories are caricatures. But they are not bad as a way of explaining the dimensions of the scientific personality. There is some overlap between the intuitive and the humanistic type. But these descriptions are pretty good." Scientist B: "I think these descriptions capture some pretty important relevant characteristics. They may be extremes but then I've seen a lot of these extremes. In this sense, they seem pretty realistic, maybe too realistic."

9 About two thirds of the way through the study of the moon scientists, I did a small study of some nine psychologists, using many of the same exercises for comparison. This is not the place to enter into a full discussion between the two groups. One thing however stands out. Unlike the moon scientists the psychologists had relatively little difficulty in naming someone within their own immediate discipline who came immediately to mind as a prime example of a Humanistic Scientist. Given their more immediate and more overt interest in people, this should come as no surprise. One would expect more potential Humanistic Scientists to be attracted to psychology and the social and behavioral sciences in general than to the physical sciences.

10 *See* Jung [215, p. 58].

11 It will be recalled that there were two related systems of types, one system in effect being a more refined subset of the other. Gross type I includes refined types 6 and 4; II includes 5 and 3; and III includes 2 and 1. Type Is were designated as theoreticians; type IIs as combiners of theory and experiment; type IIIs as pure experimentalists. The three scientists we have previously designated as $S_I{}^3$, $S_I{}^2$, and $S_I{}^1$ composed the class of type 6 scientists. Their prime distinguishing qualities were that of intense speculation and perceived commitment (see Chapter Four). Type 4s are also theoreticians and also speculative, although to a much lesser degree than $S_I{}^3$, $S_I{}^2$, and $S_I{}^1$. Type 5s are excellent and outstanding combiners of both theory and experiment. They have the ability to collect good data and to bring the necessary theoretical equipment to bear to give their data the necessary and broad interpretation it requires. Type 3s share this quality although to a lesser degree. Type 2 and 1 are primarily experimentalists of the sort who collect good data but with little or rare interpretation. They are reluctant to extrapolate beyond their data. They eschew speculation in general.

12 See Wartofsky [457] for an interesting discussion on the relation between science and metaphysics.

13 It is feeling in the Jungian sense that is being discussed. It is not true that scientists do not have feelings or emotions — of course they do. Indeed, the interview materials of Chapter Three indicate just how deeply scientists feel about their fellows and even about science itself. While obviously important, feeling and emotion in this sense are not what Jung means by feeling. As Cowan has put it: "[The] law is highly skilled in making value judgments or as I shall call them here for sake of greater precision, *feeling-value judgments*. By *feeling* I very definitely do not mean emotion, because emotion colors all mental states and functions. I mean the process

284

by which the distinctive worth of an individual is brought into view — the focusing or concentrating on a special object, or the selecting of one among a group of alternatives (decision-making). This mental function is the opposite of the function by which the human mind sees similarities and generalizes from them" [100, p. 1066].

Cowan is referring to the fact that the decision or thought process that are used in morality, ethics, law, the arts are very different from those which are used in science. It is feeling in this sense which Jung is talking about.

When I say that feeling is the psychological function that is most conspicuously absent from science, I mean that it is the function which has been the most excluded from it, the least developed within it. This exclusion of feeling is clearly seen in such traditional expressions of sentiment as "emotions have no place in science," "scientific statements are value-free," "values have no place in science," "science is the cold test of rational reason," and so on. Again as Cowan has put it: "This highly specialized art [of the scientist's method] is jealous of the scientist's energy. It entrenches on his personal life. It tempts him to neglect all aspects of his work save those that conform to the scientific ideal. These are conditions of extreme dedication. But this way of life exacts its toll. Collectively, it bears most heavily on the *feeling life of the scientist*. I believe that it is here that the scientist must make his greatest sacrifice. For example, he is often told that the power which science creates is impersonal, nonpolitical, amoral. We will not stop to debate this issue. That it can even be raised is the significant point. For another example, the scientist's expertise in the art of generalization subjects him to the risk of generalizing ordinary human relations, thus killing off the human sentiment that calls for individuation, the making unique of the relatedness that all human beings seek" [106, p. 966]. It is thus not that scientists do not have feelings or feeling but that the realm of feeling is developed strictly apart from the sphere of the scientist's working life. See Feyerabend [132, pp. 95—96] for a particularly powerful expression on this point.

Chapter 7

1 As Scheffler has put it, "That the ideal of objectivity has been fundamental to science is beyond question. The philosophical task is to assess and interpret this ideal: to ask how, if at all, objectivity is possible" [391, p. v].

2 For a precise explication of the concept of *systems separability*, see Churchman [72]. Roughly speaking, two systems are *separable* if and only if the measures of performance of one (e.g., the determination of its properties) are independent of the measures of performance of the other. Note what systems separability implies — that the two systems or activities can be designed or performed independently of one another. This is why it is so important to examine the orthodox view of scientific theories. To those of us who have to spend hours designing theoretical instru-

ments to collect all of our supposedly theory-free data, the orthodox view of science is not only naive but actually downright dangerous. It promotes a false separation among aspects of science that can only be understood conjointly.

3 Kuhn [247, 248] has pointed out how much (not all) of the objective agreement in science is artifactual. Much of the agreement arises, not because it is inherent in the structure of physical reality itself but because the scientist has been trained to look for agreement and objectivity in nature. Kuhn points out that the majority of a scientist's education consists of solving textbook (highly idealized) problems for which there exists a commonly agreed upon formulation and a single answer. According to Kuhn, agreement runs so deep that scientific history is even rewritten to conform with it. Thus, it is not that there is no disagreement in science but that the history of science is subtly rewritten (especially, in the prime vehicle of scientific education, the textbook) to suppress the conflict. The impression is given that the history of science is one continual line marked by cumulative progress. According to Kuhn, the very process of scientific education is itself responsible to a large degree for the widespread currency of the Storybook image of science.

Kuhn has also analyzed [247, 248] how the textbook is responsible for a number of other false notions regarding the nature of science. For instance, the notion that scientific laws are the cumulative assemblage and pasting together of large numbers of facts is philosophically dubious. In addition, apparently the notion is not even supported by reference to actual historical cases of scientific discovery. If one examines actual case histories, one finds intuitive, theoretical notions leading to the development of more rigorous formal theories, not necessarily the formal theory as the result of the patient accumulation of neutral facts. One of the reasons for this is that many of the supporting empirical facts could only be collected long after the theory had been formulated. According to Kuhn, at the time when many of the great theories of the physical sciences were formulated, we had neither the empirical sophistication nor necessary apparatus to test the theories. The empirical confirmation of many of our most cherished theories in the physical sciences took more than one hundred years [245].

4 I would agree with Popper [349, p. 134] on at least the following: Sometimes data guides and leads the explicit and formal development of theory, but in all cases one cannot collect scientific data without having invoked at least some notion of theory, if only implicitly, if only in the sense of having made an intuitive judgment of what is relevant to collect, observe, etc. In the language of this essay, Lockeans have unconsciously presupposed *some* Leibnizian IS in order to collect their data. Likewise, every Leibnizian system, no matter how far removed it seems from the realm of human experience, has presupposed the existence of some Lockean system in order to produce its formal propositions. Ever since Gödel we have been unable to say that the theorems of a system are merely contained in the primitives of a system. Every formal system is dependent on at least one undeniable experiential element for its operation — its guiding human manipulator, who discovers the theorems of the system. See Ravetz [357] on this last point.

It is also important to note that the presumption of a theory in order to engage in the act of observation does not make that theory true. Neither does it make the data collected under that theory strictly determined by that theory. If the theory were strictly true, if our presumptions were met in reality, then the data collected under the theory's presumption would agree exactly with that predicted by theory. The fact that we have to use theories as guides in order to be able to collect data does not mean that the world obeys our theories. Theories influence what we set ourselves to find, and without them we could find nothing, but they do not exactly determine what we shall find. These remarks are intended partly to meet some of Hesse's objections to the possible circularity of this whole procedure [184, p. 62].

5 Note that, in the typical behavioral science experiment, it is only this one characteristic of the entire set of IS features that usually constitutes the stimulus. Note that, under our formulation, enumerable ways exist to constitute (form) a stimulus or a response. For example, any or all of the four features can constitute the stimulus or response. Thus, the total number of ways that one can form stimuli and responses is $\binom{4}{1} + \binom{4}{2} + \binom{4}{3} + \binom{4}{4} = 2^4 - 1 = 15$.

6 Thus, the stimuli are already somewhat more complex to begin with than in the pure Leibnizian case, although the elements and operators can be extremely simple in any particular experimental design.

7 The purpose of this last step is to allow a decision maker to examine consciously the ethical (or esthetic) consequences of his policies by subjecting them to an intensive dialectical cross-examination by opposing moral or ethical viewpoints. If agreement is the decision rule of ordinary, or in Kuhn's [247, 248] terms, *normal*, science, then in Churchman's terms, disagreement is the decision rule of ethics. The one thing men do not universally agree on is what "ought to be." And so if we are ever to have a "science of ethics" [77], it will have to be based on a process of decision-making that explicitly allows for conflict and debate. Thus, in order to derive the ethical implications of any technical or scientific model, a dialectical mode of examining (or testing) models is explicitly incorporated. This constitutes a minimal condition for ethical evaluation. It obviously does not insure that the evaluation will be sufficent. There may be no known sufficient conditions at this time.

8 For some examples of the spirit or flavor of this approach, see Popper [347, p. 44 and 46] and Scheffler [391, p. 1].

9 This is not to say that all versions and statements of this position are for the complete elimination of emotion. Scheffler, for example, freely acknowledges that the idea of the cold and aloof scientist is a myth. However, the point seems to be that the scientist's beliefs and emotions are ultimately subject to some prime controls, among which are those stemming from the attitudes of impartiality and detachment. I too favor controls. However I differ from Scheffler in that I believe that one of the prime controls on the emotions and beliefs of one scientist are the emotions and beliefs of other scientists. In other words, I believe that the controls are not only rational or cognitive but affective as well. See Scheffler [391, p. 2].

10 Churchman and Ackoff remark on Steven's conception as follows: "Ac-

cording to this account, the criterion of the directly observable depends on the mutual agreement of normal men. The determination of such mutual agreement is, of course, a very complex problem. The determination of what constitutes normality is also no simple problem. As long as there is any error possible in our determination of normality or agreement, there would be error involved in our determination of what is directly observable. Consequently, in this operationist approach,... the isolation of the directly observable appears to represent an ideal limit to a certain kind of scientific investigation. It seems to be an ideal that is sought, rather than a real that is found" [81, pp. 460—461].

11 In Singer the reflective observer is known as the *reflective onlooker*. His purpose is to explicate those properties of the subject which the subject cannot observe of himself. Thus, Singer grants that whatever one learns is based on experience, but he also notes that one must possess specific characteristic prequisites that will make learning from experience possible. Singer notes that the experience — i.e., what is learned — belongs to the subject. However, the knowledge of (understanding of) what is learned often belongs to another — i.e., to the reflective onlooker. Thus, the basis for Singer's maxim: *"Every experiencing mind presupposes a reflective mind* [240, p. 82]." In a previous paper, I have in terms of this notion, shown why the argument of solipsism is untenable [313]. In brief, the argument is that a mind without knowledge of others would be without knowledge of itself, because the ability to have awareness of one's own states is not developed in isolation from other minds.

12 Thus, for example, from theoretical considerations one would expect Jung's thinking type to have a preference for the Leibnizian IS; the sensation type, for the Lockean IS; the intuition-thinking type for the Kantian IS, the pure intuition type for the Hegelian IS, intuitive-feeling and intuitive-thinking types for the Churchmanian-Singerian IS.

Chapter 8

1 James Fixx put it this way in a recent *Saturday Review* article entitled "The Game Game": "Huizinga [*in Homo Ludens: A Study of the Play Element in Culture*] argues that games are in no way synonymous with trivia. Rather he says, 'We... call the category *play* one of the most fundamental in life.' Play, he thinks, does not merely arise out of culture, but is an integral, inseparable part of culture. If that is so, then a society's games, and its attitudes toward those games, may tell a great deal about its people" [136, p. 64].

2 Thus, any one of a number of other metaphors could be used to characterize science. For example, C. West Churchman in *Challenge To Reason* [70] develops the theme that science is a form, albeit a very specialized one, of storytelling. The question Churchman poses is. "How fares science as a way of telling the story of nature, social as well as physical?"

3 Later on, I give a broader definition of a game. However, even at this point it can be noted that one of the most distinctive features of games is

288

that of breaking the rules. One form of this is cheating. One can still be said to be playing a game if he uses means which are not permitted by the rules of the game as long as his violation of the rules is not "too much" or does not take place "too often." Indeed a certain amount of cheating may even be necessary to keep the excitement of the game high, and particularly if it helps to expose some inevitable ambiguities that are present in the structure of every game.

4 It is not my purpose to discuss each of these aspects in detail in this chapter. Rather, the metaphor of a game is intended to emphasize that the various disciplines have traditionally claimed various aspects of The Game of Science as their own. Thus, for example, psychology has traditionally studied the "individual game players"; sociology, the "groups of players"; the logic of science, the "rules". Primarily the metaphor of a game is intended to emphasize that The Game of Science is fundamentally a human creation, not nature's.

5 Consider Churchman on this point: "The third level [of standardization of data] consists of adjusting all or almost all data to standards by means of laws that enable one to say: *if* report R_1 was made at time t_1 in circumstance z_1 by a person having properties w_{11}, w_{12}, and so on, then report R_0 would have been made at time t_0 in circumstance z_0 by a person having properties w_{01}, w_{02}, and so on. The last level most clearly approximates the function of measurement, because it permits the broadest use of the information" [77, p. 121].

6 I would especially stress the word *dependent* in "presupposition-dependent." I do not believe that presuppositions determine, but rather influence what an observer observes. In this sense, I would soften Shapere's characterization of a "presupposition view of scientific theories" [396].

7 A number of the respondents put it in nearly exactly these words in discussing some of their peers.

8 There are also "liberals" and "conservatives"; the number of types is by no means restricted to merely three.

9 Martin himself comes close to stating this. Consider the following: "For Feyerabend scientists have a *choice* [emphasis added]: they can *choose* [emphasis added] to make their theories incommensurable with one another or they can *choose* [emphasis added] not to make them incommensurable [276, pp. 1–2]." The point is whether a particular individual can or cannot make the choice is at least partially dependent on his personality, his psychology.

10 In another sense it can be said that Kuhn forgot one of his own earlier pieces which contained precisely the psychology he needed. I refer to Kuhn's paper, "The Essential Tension: Tradition and Innovation in Scientific Research" [244], in which he commented that the progress (advancement) of science depended on a tug-of-war between two fundamental types: the innovator and the traditionalist. The one was dedicated to breaking new ground; the other, to preserving the past.

11 And again, as Ackoff stresses, they are not independent. Thus, to a pragmatist such as myself, even if one's goal is to increase the truth content of our theories, this is not necessarily independent of increasing the social good, because for pragmatism one of the measures of truth is in terms of increased social benefits.

289

12 In many senses it is to honor its creative and playlike aspects.

13 It is beyond the scope of this book to argue why the results of this study do not necessarily lead to relativism. The outlines of such an argument have already been given by Churchman [79] and Toulmin [446]. An exceedingly brief response is that relativism does not inevitably follow just because the system is crucially dependent on a number of human and even irrational elements. Relativism would be the inevitable response if and only if these human and irrational elements were not susceptible to systematic study of any type. One of the points of this study has been to show that the psychology of scientists is not a random, aimless phenomenon. Irrationality is as patterned and structured as any of the phenomena of nature. It may not be lawlike in the way that philosophers usually think of laws, but it is not unstructured. Just because The Game is fundamentally dependent on the psychology of its players does not mean that the course of play is ruled by mob psychology as some [252] are wont to put it. I find truly incredible the leap from the fact that scientists have a psychology to the supposition that therefore mob psychology rules the game.

14 For example, although I might agree with Theodore Roszak that there is much in science that justifies characterizing science as *"The Myth of Objective Consciousness* [380]," I do not see the counter-culture as the promised antidote. If science embodies the myth of objective consciousness, then too much of Roszak's arguments embody *the myth of subjective consciousness.* If I do not buy science or thinking as *the* antidote to all our problems, then I do not buy the counter-culture or feeling as the answer either. As a disciple of Jung, I am interested in promoting a reconciliation between thinking and feeling, not in promoting their further separation.

15 It is a cliché because there never was a time that this could not be said. Surely there were scientists before the term was invented. The official term is only a couple of hundred years old at best anyway. If we define a scientist as one who embodies a particular set of psychological attributes, as one who views nature in prescribed ways, then there is no reason to believe that we have not had scientists with us as far back as the human race began to develop differentiated personality structures.

16 I find it amusing that, in the few times I have talked about this study before public groups, there have always been those few scientists in the audience who are insistent that if I had studied such and such a group or their particular brand of science (discipline), I would not have found such outrageous behavior. What is so amusing is that each group of scientists is so willing to believe that the other group is outrageous, not themselves, of course!

Nevertheless I would be the first to admit that the results of this study might not apply in their entirety to every group of scientists, but only in the sense that not every group might exhibit the effects reported here with the same strength and degree of clarity. Thus, for example, although I admit that the particular group I studied may be more aggressive than the average, I would still expect, because of the prior literature cited on the psychology of scientists (see Chapter Four), every scientific group to exhibit noticeable degrees of aggressiveness. Obviously, though, this study

290

should be replicated across different disciplines, cultures, laboratories, universities, etc. I hope my results encourage others to do this.

17 *The Game is also, in the end, supremely governed by metaphysical considerations.* For example, in order for Maxwell to "show that the aim of seeking theories of both increasing explanatory power (or simplicity) and verisimilitude is a rational aim to adopt" [284, p. 149], he must introduce a metaphysical principle, "the assumption that the thesis of structural simplicity is true" [284, p. 149]. I have no objection to introducing methaphysical principles. Indeed, I think they should be consciously introduced so that we can explicitly challenge and question the principles we wish to introduce. But I find it amusing that Maxwell solves the demarcation problem of distinguishing between scientific and nonscientific theories by introducing a nonscientific postulate of his own, i.e, the metaphysical principle of structural simplicity.

Bibliography

1. ————, *American Men and Women of Science: A Biographical Directory*. New York: Jacques Cattell Press, R.R. Bowker, 1972.
2. ————, *American Men of Science: A Biographical Directory*. New York: Jacques Cattell Press, R.R. Bowker, 1967.
3. ————, *Proceedings of the Apollo 11 Lunar Science Conference*. *Science*, January 30, 1970.
4. ————, *Proceedings of the Second Lunar Science Conference*. Cambridge: Massachusetts Institute of Technology Press, 1971.
5. ————, *Proceedings of the Third Lunar Science Conference, Vol. I: Mineralogy and Petrology*, Elbert A. King (Ed.) *Vol. II: Chemical and Isotope Analyses/Organic Chemistry*, Dieter Heymann (Ed.) *Vol. III: Physical Properties*, David R. Criswell (Ed.) Cambridge: Massachusetts Institute of Technology Press, 1972.
6. ————, *Science Citation Index*. Philadelphia: Institute for Scientific Information, 1964–1971.
7. ————, *Tektites: A Bibliography*. Redstone Arsenal, Ala.: Redstone Scientific Information Center, January 15, 1964.
8. ————, *Theory of Continental Drift*. Tulsa, Okla.: American Association of Petroleum Geologists, 1928.
9. Abelson, Robert P., *et al.* (Eds.) *Theories of Cognitive Consistency: A Sourcebook*. Chicago: Rand McNally, 1968.
10. Abramson, E., *et al.* "Social Power and Commitment: A Theoretical Statement." *American Sociological Review*, Feb. 1958, *23* (1), 15–22.
11. Ackoff, Russell L. *The Design of Social Research*. Chicago: University of Chicago Press, 1953.
12. Ackoff, Russell L. *Scientific Method, Optimizing Applied Research Decisions*. New York: John Wiley, 1962.
13. Ackoff, Russell L. "Toward An Idealized University." *Management Science*, Dec. 1970, *15* (4), B–121 to B–131.
14. Ackoff, Russell L. "Towards A Behavioral Theory Of Communication," *Management Science*, 1958, *4*, 218–234.
15. Ackoff, Russell L. "Towards a System of Systems Concepts." *Management Science*, July 1971, *17* (11), 661–671.
16. Ackoff, Russell L., and Emery, Fred. *On Purposeful Systems*. Chicago: Aldine-Atherton, 1972.
17. Agassi, Joseph. *Towards an Historiography of Science*. The Hague: Mouton, 1963.
18. Ahrens, L.H. *Distribution of the Elements in Our Planet*. New York: McGraw-Hill, 1965.

19. Albritton, Claude Carrol (Ed.) *The Fabric of Geology*. Reading, Mass.: Addison-Wesley, 1963.
20. Allen, Allen D. "Scientific Versus Judicial Fact Finding in the United States." *IEEE Trans. on Systems, Man, and Cybernetics*, Sept. 1972, 548—550.
21. Allport, G.W., Vernon, P.E., and Lindzey, G. *Study of Values*. Boston: Houghton-Mifflin, 1931.
22. Anderson, Don L., and Hanks, Thomas, C. "Is the Moon Hot or Cold?" *Science*, Dec. 22, 1972, *178* (4067), pp. 1245—1249.
23. Anderson, Harold H., and Anderson, Gladys L. *An Introduction to Projective Techniques*. Englewood Cliffs, N.J.: Prentice Hall, 1951.
24. Apollo 15 Preliminary Examination Team. "The Apollo 15 Lunar Samples: A Preliminary Description." *Science*, Jan. 28, 1972, *175* (4020), pp. 363—375.
25. Apollo 16 Preliminary Examination Team. "The Apollo 16 Lunar Samples: A Petrographic and Chemical Description." *Science*, Jan. 5, 1973, *179* (4068), pp. 23—24.
26. Asch, S.E. "Forming Impressions of Personality." *Journal of Abnormal and Social Psychology*, 1946, *41*, 258—290.
27. Bain, Read. "The Scientist and His Values." *Social Forces*, Dec. 1952, *31* (2), 106—109.
28. Bakan, David. *On Method: Toward a Reconstruction of Psychological Investigation*. San Francisco: Jossey-Bass, 1967.
29. Baldwin, Ralph B. *A Fundamental Survey of the Moon*. New York: McGraw Hill, 1965.
30. Baldwin, Ralph B. *The Measure of the Moon*. Chicago: University of Chicago, 1963.
31. Baldwin, Ralph B. "Summary of Arguments for a Hot Moon." *Science*, Dec. 18, 1970, *170* (3964), 1297—1300.
32. Barber, Bernard. "Resistance by Scientists to Scientific Discovery." In Bernard Barber and Walter Hirsch (Eds.), *The Sociology of Science*. New York: Free Press, 1962.
33. Barber, Bernard. *Science and the Social Order*, New York: Collier, 1962.
34. Barber, Bernard, and Hirsch, Walter (Eds.) *The Sociology of Science*. New York: Free Press, 1962.
35. Barber, Theodore Xenophon, and Silver, Maurice J. "Fact, Fiction, and the Experimenter Bias Effect." *Psychological Bulletin*, Dec. 1968, *70* (6), 1—29.
36. Barrett, Robert. "On the Conclusive Falsification of Scientific Hypotheses," *Philosophy of Science*, Dec. 1969, *36* (4), 363—374.
37. Becker, Ernest. *The Structure of Evil, An Essay on the Unification of the Science of Man*. New York: George Braziller, 1968.
38. Becker, Howard S. "Notes on the Concept of Commitment." *American Journal of Sociology*, July 1960, *66* (1), 32—40.
39. Ben-David, Joseph. *The Scientist's Role in Society — A Comparative Study*. Englewood Cliffs, N.J.: Prentice-Hall, 1971.
40. Blackburn, Thomas R. "Sensuous-Intellectual Complementarity in Science." *Science*, June 4, 1971, *172* (3987), 1003—1007.
41. Blackwell, Richard J. *Discovery in the Physical Sciences*. Notre Dame, Ind.: University of Notre Dame Press, 1969.

42. Blanco, V.M., and McCuskey, S.W. *Basic Physics of the Solar System*. Reading, Mass.: Addison-Wesley, 1961.
43. Blankenship, L. Vaughn. "Public Administration and the Challenge to Reason." In Dwight Waldo (Ed.), *Public Administration in a Time of Turbulence*. London: Chandler, 1971.
44. Blankenship, L. Vaughn. "Review of 'The Decision To Go To The Moon'." *Science*, July 23, 1971, *173* (3994), 317—318.
45. Blankenship, L. Vaughn. "The Scientist As 'Apolitical' Man." Paper presented at the meeting of the Northeastern Political Science Association. Philadelphia, Nov. 12—14, 1970.
46. Bloor, David. "Two Paradigms for Scientific Knowledge." *Science Studies*, 1971, *1*, 101—115.
47. Boalt, Gunnar. *The Sociology of Research*. Carbondale, Ill.: Southern Illinois University Press, 1969.
48. Bohm, David. "Fragmentation in Science and in Society." *Impact of Science on Society*, 1970, *20* (2), 159—169.
49. Boring, Edwin G. "Cognitive Dissonance: Its Use in Science." *Science*, Aug. 14, 1964, *145*, 680—685.
50. Brager, George. "Commitment and Conflict in Normative Organization." *American Sociological Review*, 1969, *34*, 482—491.
51. Brain, Walter Russell. "Science and Antiscience." *Science*, April 9, 1965, *148*, 192—198.
52. Braithwaite, Richard Bevan. *Scientific Explanation, A Study of the Function of Theory, Probability and Law in Science*. Cambridge, England: Cambridge University Press, 1953.
53. Brown, Norman O. *Life Against Death, The Psychoanalytical Meaning of History*. New York: Vintage, 1959.
54. Brownhill, R.J. "Scientific Ethics and The Community, The Applicability of Polanyi's Concept of Ethics in The Scientific Community to The Community as a Whole." *Inquiry*, *11*, 243—248.
55. Bruner, Jerome S., and Goodman, Cecile C. "Value and Need as Organizing Factors in Perception." *Journal of Abnormal and Social Psychology*, 1947, *42*, 33—44.
56. Bunge, Mario. *Intuition and Science*. Englewood Cliffs, N.J.: Prentice-Hall, 1962.
57. Burmester, Mary Alice. "Behavior Involved in the Critical Aspects of Scientific Thinking." *Science Education*, Dec. 1952, *36* (5) 259—263.
58. Campbell, Joseph. *The Hero With A Thousand Faces*, New York: Meridan, 1956.
59. Campbell, Joseph. "The Historical Development of Mythology." In Henry A. Murray (Ed.), *Myth and Mythmaking*. New York: George Braziller, 1960.
60. Campbell, Norman. *What Is Science?* New York: Dover, 1952.
61. Carter, Launor F., and Schooler, Kermit. "Value, Need, and Other Factors in Perception." *Psychological Review*, 1949, *56*, 200—207.
62. Cartter, Allan M. *An Assessment of Quality in Graduate Education*. Washington, D.C.: American Council on Education, 1966.
63. Cartwright, Dorwin, and Harary, Frank. "Structural Balance: A Generalization of Heider's Theory." *Psychological Review*, 1956, *63*, 277—293.
64. Cattell, R.B., and Drevdahl, J.E. "A Comparison of the Personality Profile

(16 p.f.) of Eminent Researchers with That of Eminent Teachers and Administrators, and of The General Population." *British Journal of Psychology*, 1955, *46*, 248—261.

65. Cattell, R.B. *Personality and Motivation: Structure and Measurements.* Yonkers-on-Hudson, N.Y.: World Books, 1957.

66. Caws, Peter. "The Structure of Discovery." *Science*, Dec. 12, 1969, *166*, 1375—1380.

67. Chamberlin, T.C. "The Method of Multiple Working Hypotheses." *Science*, May 7, 1965, *148*, 754—759.

68. Chapman, Dean R. "Australasian Tektite Geographic Pattern, Crater and Ray of Origin, and Theory of Tektite Events." *Journal of Geophysical Research*, Sept. 10, 1971, *76* (26), 6309—6338.

69. Churchman, C. West. "The Artificiality of Science." *Contemporary Psychology*, June, 1970, *15* (6), 385—386.

70. Churchman, C. West. *Challenge to Reason.* New York: McGraw-Hill, 1968.

71. Churchman, C. West. "Concepts Without Primitives." *Philosophy of Science*, Oct. 1953, *20* (4), 257—265.

72. Churchman, C. West. *The Design of Inquiring Systems.* New York: Basic Books, 1971.

73. Churchman, C. West. "Kant — A Decision Theorist?" *Theory and Decision*, Oct. 1970, *1* (1), 107—116.

74. Churchman, C. West. "On Whole Systems." Internal Working Paper 31, May 1965b, Space Sciences Laboratory, University of California, Berkeley, California.

75. Churchman, C. West. "Operations Research as a Profession." *Management Science*, Oct. 1970, *17* (2), B-37 to B-53.

76. Churchman, C. West. "The Philosophy of Experimentation." In Oscar Kempthorne, et al. (Eds.) *Statistics and Mathematics in Biology.* Ames, Iowa: Iowa State College Press, 1954.

77. Churchman, C. West. *Prediction and Optimal Decision: Philosophical Issues of a Science of Values.* Englewood Cliffs, N.J.: Prentice-Hall, 1961.

78. Churchman, C. West. "Research on Research, A Philosophical Discussion of Self-Reflection." Internal Working Paper 69, August 1967, Space Sciences Laboratory, University of California, Berkeley, California.

79. Churchman, C. West. *Theory of Experimental Inference.* New York: Macmillan, 1948.

80. Churchman, C. West, and Ackoff, Russell L. "An Experimental Measure of Personality." *Philosophy of Science*, 1947, *14*, 304—332.

81. Churchman, C. West, and Ackoff, Russell L. *Methods of Inquiry, An Introduction to Philosophy and Scientific Method.* St. Louis: Educational Publishers, 1950.

82. Churchman, C. West; Ackoff, Russell L.; and Wax, Murray (Eds.) *Measurement of Consumer Interest.* Philadelphia: University of Pennsylvania Press, 1947.

83. Churchman, C. West, and Ratoosh, Philburn (Eds.) *Measurement, Definitions and Theories.* New York: John Wiley, 1959.

84. Churchman, C. West, and Schainblatt, A.H. "On Mutual Understanding." *Management Science*, Oct. 1965, *12* (2), B-40 to B-42.

85. Cloud, Preston. "Lunar Science and Planetary History." Guest Editorial, *Science*, Sept. 18, 1970, *169* (3951), 1159.
86. Coan, Richard W. "Dimensions of Psychological Theory." *American Psychologist*, 1968, *23*, 715—722.
87. Cohen, A.R. *Attitude Change and Social Influence*. New York: Basic Books, 1964.
88. Cohen, David, "Curriculum Objective: Scientific Attitudes," Paper presented at the Curriculum Studies Symposium, 43rd ANZAAS Congress, Brisbane, Australia, May 26, 1971.
89. Cohen, David, "Personality of Professionals." *Science Journal*, Jan. 1971, *7* (1), 85—87.
90. Cole, Jonathon R., and Cole, Stephen. "The Ortega Hypothesis." *Science*, Oct. 27, 1972, *178* (4059), 368—375.
91. Colodny, Robert G. *Beyond the Edge of Certainty, Essays in Contemporary Science and Philosophy*. Englewood Cliffs, N.J.: Prentice-Hall, 1965.
92. Colodny, Robert G. (Ed.) *Mind and Cosmos, Essays in Contemporary Science and Philosophy*. Pittsburgh: University of Pittsburgh Press, 1966.
93. Colodny, Robert G. *The Nature and Function of Scientific Theories, Essays in Contemporary Science and Philosophy*. Pittsburgh: University of Pittsburgh Press, 1970.
94. Conant, James B. "Scientific Principles and Moral Conduct." *American Scientist*, 1967, *55*(3), 311—328.
95. Cooley, William W., and Lohnes, Paul R. *Multivariate Procedures for the Behavioral Sciences*. New York: John Wiley, 1962.
96. Cooper, Henry S.F. Jr. *Apollo on the Moon*. New York: Dial Press, 1969.
97. Cooper, Henry S.F. Jr. *Moon Rocks*, New York: Dial Press, 1970.
98. Coser, Lewis. *The Functions of Social Conflict*. New York: Free Press, 1956.
99. Cousins, Norman. "Lunar Meditations." Editorial, *Saturday Review*, August 14, 1971, p. 20.
100. Cowan, Thomas A. "Decision Theory in Law, Science and Technology." *Science*, June 7, 1963, *140*, 1065—1075.
101. Cowan, Thomas A. "Experience and Experiment." *Philosophy of Science*, April 1959, *26* (2), 77—83.
102. Cowan, Thomas A. "The Game of Science." Paper presented at the Symposium on Philosophy of Science (Section L), sponsored by the American Association for the Advancement of Science, Berkeley, California, Dec. 1965.
103. Cowan, Thomas A. "A Model for Jurisprudential Investigation, the Lawyer as an Emerging Force in Modern Jurisprudence." Internal Working Paper No. 60, Space Sciences Laboratory, Social Sciences Project, University of California, Berkeley, California, March 1967.
104. Cowan, Thomas A. "Non-Rationality in Decision Theory." Internal Working Paper 36, Space Sciences Laboratory, Social Sciences Project, University of California, Berkeley, California, Nov. 1965.
105. Cowan, Thomas A. "A Note on Equilibrium Systems from a Dialectical (Tensional) Point of View," Internal Working Paper No. 50, Space Sci-

ences Laboratory, Social Sciences Project, University of California, Berkeley, California, Oct. 1966.

106. Cowan, Thomas A. "Paradoxes of Science Administration." *Science,* Sept. 15, 1972, *177,* 964—966.

107. Cowan, Thomas A. "A Postulate Set for Experimental Jurisprudence." *Philosophy of Science,* Jan. 1951, *18* (1), 1—15.

108. Crane, Diana. "The Gatekeepers of Science: Some Factors Affecting the Selection of Articles for Scientific Journals." *The American Sociologist,* 1967, *2,* 195—201.

109. Crane, Diana. "Social Structure in a Group of Scientists: A Test of the 'Invisible College' Hypothesis." *American Sociological Review,* June 1969, *34* (3), 335—352.

110. Crockett, Campbell. "Misunderstanding One Another." In Sidney Hook (Ed.) *Psychoanalysis, Scientific Method, and Philosophy.* New York: New York University Press, 1959.

111. Cronbach, Lee J., and Gleser, Goldine C. "Assessing Similarity Between Profiles." *The Psychological Bulletin,* 1953, *66* (6), 456—473.

112. Daniels, George H. "The Pure Science Ideal and Democratic Culture." *Science,* June 30, 1967, *156,* 1699—1705.

113. Davis, Ira C. "The Measurement of Scientific Attitudes." *Science Education, 19* (3), 117—122.

114. Davis, James A. "Structural Balance, Mechanical Solidarity, and Interpersonal Relations," *American Journal of Sociology,* 1963, *68,* 444—462.

115. Davis, James, and Leinhardt, Samuel. "The Structure of Positive Interpersonal Relations in Small Groups." Paper read at The American Sociological Association, Boston, 1968.

116. Davis, Murray S. "That's Interesting!: Towards a Phenomenology of Sociology and a Sociology of Phenomenology." *Philosophy of The Social Sciences,* Dec. 1971, *1* (4), 309—344.

117. de Beauvoir, Simone. *The Second Sex.* Trans and ed. H.M. Parshley. New York: Bantam, 1961.

118. DeBurger, Robert A., and Donahoe, John W. "Relationships Between the Meanings of Verbal Stimuli and Their Associative Responses." In James G. Snider and Charles E. Osgood (Eds.) *Semantic Differential Technique.* Chicago: Aldine, 1969.

119. Dewey, John. *The Quest For Certainty.* New York: G.P. Putnam, 1960.

120. Diederich, Paul B. "Components of the Scientific Attitude." *Science Teacher,* 1967, *34* (1), 23—24.

121. Diesing, Paul. "Subjectivity and Objectivity in the Social Sciences." *Philosophy of the Social Sciences.* June 1972, *2* (2), 147—165.

122. Duhem, Pierre. *The Aim and Structure of Physical Theory.* Princeton, N.J.: Princeton University Press, 1954.

123. Ebel, Robert L. "What Is the Scientific Attitude?" *Science Education,* 1938, *22,* 1—5, 75—81.

124. Edgington, Eugene S. "Statistical Inference and Nonrandom Samples." *Psychological Bulletin,* 1966, *66,* 485—487.

125. Edwards, Allen L. *Experimental Design in Psychological Research.* New York: Holt, Rinehart, and Winston, 1960.

126. Eglinton, Geoffrey; Maxwell, James R.; and Pillinger, Colin T. "The

Carbon Chemistry of The Moon." *Scientific American,* Oct. 1972, *227* (4), 80—90.

127. Eicher, Don L. *Geologic Time.* Englewood Cliffs, N.J.: Prentice-Hall, 1968.

128. Eiduson, Bernice T. *Scientists: Their Psychological World.* New York: Basic Books, 1962.

129. Feigl, Herbert. "The 'Orthodox' View of Theories: Remarks in Defense as Well as Critique." In Michael Radner and Stephen Winokur (Eds.) *Analyses of Theories and Methods of Physics and Psychology, Minnesota Studies in the Philosophy of Science, Vol. IV.* Minneapolis: University of Minnesota Press, 1970.

130. Feigl, Herbert, and Brodbeck, May. *Readings in The Philosophy of Science.* New York: Meridian Books, 1960.

131. Ferdinand, Theodore N. "Personality and Career Aspirations Among Young Technologists." *Human Relations, 22* (2), 121—135.

132. Feyerabend, Paul K. "Against Method: Outline of an Anarchistic Theory of Knowledge." In Michael Radner and Stephen Winokur (Eds.) *Analyses of Theories and Methods of Physics and Psychology, Minnesota Studies in the Philosophy of Science, Vol. IV.* Minneapolis: University of Minnesota Press, 1970.

133. Feyerabend, Paul K. "On the Improvement of the Sciences and Arts, and the Possible Identity of the Two." In Robert S. Cohen and Marx W. Wartofsky (Eds.) *Boston Studies in the Philosophy of Science.* Dordrecht, Holland: D. Reidel, 1967.

134. Feyerabend, Paul K. "Problems of Empiricism." In Robert G. Colodny (Ed.) *Beyond the Edge of Certainty,* Englewood Cliffs, N.J.: Prentice Hall, 1965.

135. Feyerabend, Paul K. "Problems of Empiricism, Part II" In Robert G. Colodny (Ed.) *The Nature and Function of Scientific Theories.* Pittsburgh: University of Pittsburgh Press, 1970.

136. Fixx, James F. "The Game Game." *Saturday Review of Education,* Jan. 1973. pp. 62—66.

137. Foss, Laurence. "Art as Cognitive: Beyond Scientific Realism." *Philosophy of Science,* June 1971, *38* (2), 234—250.

138. Frey, Fred A.; Spooner, Charles M.; and Baedecker, Philip A. "Microtektites and Tektites: A Chemical Comparison." *Science,* Nov. 20, 1970, *170* (3960), pp. 845—847.

139. Friedman, Neil. *The Social Nature of Psychological Research: The Psychological Experiment as a Social Interaction.* New York: Basic Books, 1967.

140. Garner, Wendell, R. *Uncertainty and Structure as Psychological Concepts.* New York: John Wiley, 1962.

141. Garvey, William D., and Griffith, Belver C. "Scientific Communication as a Social System." *Science,* Sept. 1, 1967, *157,* pp. 1011—1016.

142. Garvey, William D.; Lin, Nan; and Nelson, Carnot E. "Communication in The Physical and The Social Sciences." *Science,* Dec. 11, 1970, *170,* 1166—1173.

143. Gaston, Jerry. "Secretiveness and Competition for Priority of Discovery in Physics." *Minerva,* Oct. 1971, *9* (4), 472—492.

144. Gaston, Jerry. "Social Processes in Science." *New Scientist*, Dec. 7, 1972, *56* (823), pp. 581—583.
145. Gerald, Harold B. "Deviation, Conformity, and Commitment." In Ivan D. Steiner and Martin Fishbein (Eds.) *Current Studies in Social Psychology*. New York: Holt, Rinehart, and Winston, 1965.
146. Gillette, Robert. "The Aftermath of Apollo: Science on The Shelf." *Science*, Dec. 22, 1972, *178* (4067), pp. 1265—1268.
147. Gillispie, Charles Coulston. *The Edge of Objectivity, An Essay in the History of Scientific Ideas*. Princeton: Princeton University Press, 1960.
148. Ginzberg, Eli. *The Development of Human Resources*. New York: McGraw Hill, 1966.
149. Glaser, Barney G. "Comparative Failure in Science." *Science*, March 6, 1964, *143*, pp. 1012—1014.
150. Glaser, Barney G. "Variations in The Importance of Recognition in Scientists' Careers." *Social Problems*, 1963, *10*, 268—276.
151. Glass, Bentley. "The Ethical Basis of Science." *Science*, Dec. 3, 1965, *150*, pp. 1254—1261.
152. Goldreich, Peter. "Tides and The Earth — Moon System." *Scientific American*, April 1972, *226* (4), 42—52.
153. Gordon, T.J., and Raffensperger, M.J. "A Strategy for Planning Basic Research," *Philosophy of Science*, June 1969, *36* (2), 205—218.
154. Gouldner, Alvin W. *The Coming Crisis of Western Sociology*. New York: Avon, 1970.
155. Graham, Patricia Albjerg. "Women in Academe." *Science*, Sept. 25, 1970, *169*, pp. 1284—1290.
156. Green, Paul E., and Carmone, Frank J. "Multidimensional Scaling: An Introduction and Comparison of Nonmetric Unfolding Techniques." *Journal of Marketing Research*, Aug. 1969, New York: *6*, 330—341.
157. Greenberg, Daniel S. *The Politics of Pure Science*. New York: New American Library, 1968.
158. Greeno, James G. "Evaluation of Statistical Hypotheses Using Information Transmitted." *Philosophy of Science*, June 1970, *37* (2), 279—294.
159. Grünbaum, Adolf. "Can We Ascertain The Falsity of a Scientific Hypothesis?" In M. Mandelbaum (Ed.) *Observation and Theory in Science*. Baltimore: Johns Hopkins Press, 1971.
160. Grünbaum, Adolf. *Philosophical Problems of Space and Time*. New York: A. Knopf, 1963.
161. Guilford, J.P. *Personality*. New York: McGraw Hill, 1959.
162. Gundlach, Ralph H., and Gerum, Elizabeth. "Vocational Interests and Types of Ability." *Journal of Educational Psychology*. 1931, *22*, 505—511.
163. Haberer, J. "Politicalization in Science." *Science*, Nov. 17, 1972, *178* (4062), pp. 713—724.
164. Hagstrom, Warren O. *The Scientific Community*. New York: Basic Books, 1965.
165. Haigh, Gerard V., and Fiske, Donald W. "Corroboration of Personal Values as Selective Factors in Perception." *Journal of Abnormal and Social Psychology*, 1952, *47*, 394—398.

166. Haney, Richard E. "The Development of Scientific Attitudes." *Science Teacher*, Dec. 1964, *31* (8), 33—35.
167. Hanson, Norwood Russell. "Is There A Logic of Discovery?" In Herbert Feigl and Grover Maxwell (Eds.) *Current Issues in the Philosophy of Science.* New York: Holt, Reinhart and Winston, 1961.
168. Hanson, Norwood Russell. "Observation and Interpretation." In Sidney Morgenbesser (Ed.) *Philosophy of Science Today.* New York: Basic Books, 1967.
169. Hanson, Norwood Russell. *Patterns of Discovery.* Cambridge, England: Cambridge University Press, 1965.
170. Hanson, Norwood Russell. *Perception and Discovery, An Introduction to Scientific Inquiry.* San Francisco: Freeman, Cooper, 1969.
171. Hanson, Norwood Russell. "A Picture Theory of Theory Meaning." In Michael Radner and Stephen Winokur (Eds.) *Analyses of Theories and Methods of Physics and Psychology, Minnesota Studies in the Philosophy of Science, Vol. IV.* Minneapolis: University of Minnesota Press, 1970.
172. Harding, M. Esther. *Woman's Mysteries Ancient and Modern, A Psychological Interpretation of the Feminine Principle as Portrayed in Myth, Story, and Dreams.* New York: Putnam, 1971.
173. Harr, D. Ter, and Cameron, A.G.W. "Historical Review of Theories of the Origin of the Solar System." In Robert Jastrow and A.G.W. Cameron, (Eds.) *Origin of the Solar System.* New York: Academic Press, 1963.
174. Hartsough, Ross, and Laffal, Julius. "Content Analysis of Scientific Writings." *The Journal of General Psychology*, 1970, *83*, 193—206.
175. Hattiangadi, J.N. "Alternatives and Incommensurables: The Case of Darwin and Kelvin." *Philosophy of Science.* Dec. 1971, *38* (4), 502—507.
176. Hayek, F.A. *The Counter Revolution of Science, Studies on The Abuse of Reason.* New York: Free Press, 1955.
177. Hegel, G.W.F. *The Phenomenology of Mind.* Trans, J.B. Baillie. London: George Allen and Unwin, 1949.
178. Heider, Fritz. *The Psychology of Interpersonal Relations.* New York: John Wiley, 1958.
179. Helmer, Olaf, and Rescher, Nicholas. "On the Epistemology of the Inexact Sciences." *Management Science*, Oct. 1959, *6* (1), 25—52.
180. Hempel, Carl G. "On The 'Standard Conception' of Scientific Theories." In Michael Radner and Stephen Winokur (Eds.) *Analyses of Theories and Methods of Physics and Psychology.* Minneapolis: University of Minnesota Press, 1970.
181. Hempel, Carl G. *Philosophy of Natural Science.* Englewood Cliffs N.J.: Prentice-Hall, 1966.
182. Hess, Wilmot; Kovach, Robert; Gast, Paul W.; and Simmons, Gene. "The Exploration of the Moon." *Scientific American*, Oct. 1969, *221* (4), 54—72.
183. Hesse, Hermann, *Magister Ludi,* New York: Holt, 1949.
184. Hesse, Mary. "Is There an Independent Observation Language?" In Robert G. Colodny (Ed.) *The Nature and Function of Scientific Theories, Essays in Contemporary Science and Philosophy."* Pittsburgh: University of Pittsburgh Press, 1970.

185. Hesse, Mary B. *Models and Analogies in Science.* Notre Dame, Ind.: University of Notre Dame Press, 1966.
186. Hill, Karl (Ed.) *The Management of Scientists.* Boston: Beacon, 1964.
187. Hillman, James. *The Myth of Analysis, Three Essays in Archetypal Psychology.* Evanston, Ill.: Northwestern University Press, 1972.
188. Hinners, N.W. "The New Moon: A View." *Reviews of Geophysics and Space Physics,* Aug. 1971, *9* (3), 447—522.
189. Hirsch, Walter. *Scientists in American Society.* New York: Random House, 1968.
190. Hitt, William D. "Two Models of Man." *American Psychologist,* July 1969, *24* (7), 651—658.
191. Hoffman, R.L.; Harburg, E.; and Maier, N.R.F. "Differences and Disagreement as Factors in Creative Group Problem Solving." *Journal of Abnormal and Social Psychology,* 1962, *64* (3), 206—214.
192. Holland, Paul, and Leinhardt, Samuel. "A Method for Detecting Structure in Sociometric Data." School of Urban and Public Affairs Working Paper, Carnegie-Mellon University, Pittsburgh, Pa.
193. Hollinger, David A. "T.S. Kuhn's Theory of Science and Its Implications for History." *American Historical Review,* April 1973, *78* (2), 370—393.
194. Holmes, David S., and Appelbaum, Alan S. "Nature of Prior Experimental Experience As A Determinant of Performance in A Subsequent Experiment." *Journal of Personality and Social Psychology,* March 1970, *14* (3), 195—202.
195. Holton, Gerald. "Introduction." In Gerald Holton (Ed.) *Science and Culture, A Study of Cohesive and Disjunctive Forces.* Boston: Beacon, 1957.
196. Holton, Gerald (Ed.) *The Making of Modern Science: Biographical Studies. Daedalus, 99* (4). (Proceedings of the American Academy of Arts and Sciences, Fall, 1970).
197. Holton, Gerald. "The Metaphor of Space-Time Events in Science." *Eranos-Jahrbuch,* 1965, *34,* 33—78.
198. Holton, Gerald. "Science and New Styles of Thought." *The Graduate Journal,* Spring 1967, *7* (2), 399—422.
199. Holton, Gerald. "The Thematic Imagination in Science." In Gerald Holton (Ed.) *Science and Culture, A Study of Cohesive and Disjunctive Forces.* Boston: Beacon, 1965.
200. Hook, Sidney. "Science and Mythology in Psychoanalysis." In Sidney Hook (Ed.) *Psychoanalysis, Scientific Method and Philosophy.* New York: New York University Press, 1959.
201. Hovard, Richard B. "Theoretical Reduction: The Limits and Alternatives to Reductive Methods in Scientific Explanation." *Philosophy of the Social Sciences,* Jan. 1971, *1* (1), 83—100.
202. Hovland, C.I.; Janis, I.; and Kelley, H. *Communication and Persuasion, Psychological Studies of Opinion Change.* New Haven, Conn.: Yale University Press, 1953.
203. Hudson, Liam. *Contrary Imaginations.* New York: Schocken Books, 1966.
204. Hudson, Liam (Ed.) *The Ecology of Human Intelligence.* Harmondsworth, England: Penguin, 1970.

205. Hudson, Liam. *Frames of Mind, Ability, Perception and Self-Perception in the Arts and Sciences.* New York: Norton, 1968.
206. Hughes, Raymond M. *Report of Committee on Graduate Instruction.* Washington, D.C.: American Council on Education, 1934.
207. Humphreys, Willard C. *Anomalies and Scientific Theories.* San Francisco: Freeman, Cooper, 1968.
208. Jacobi, Jolande J. *The Psychology of C.G. Jung,* (7th ed.) New Haven, Conn.: Yale University Press, 1968.
209. James, William. *Pragmatism and Other Essays.* New York: Washington Square Press, 1963.
210. Jarrett, R.F. and Sherriffs, A.C. "Propaganda, Debate, and Impartial Presentation as Determiners of Attitude Change." *Journal of Abnormal and Social Psychology,* 1953, *48* (1), 33—41.
211. Jastrow, Robert, and Cameron, A.G.W. (Eds.) *Origin of the Solar System.* New York: Academic Press, 1963.
212. Jastrow, Robert, and Newell, Homer E. "The Space Program and The National Interest." *Foreign Affairs,* April 1972, *50* (3), 532—544.
213. Jung, C.G. "Aion." In Violet S. de Laszlo (Ed.) *Psyche and Symbol, A Selection from the Writings of C.G. Jung.* New York: Doubleday, 1958.
214. Jung, C.G. *Analytical Psychology, Its Theory and Practice.* New York: Pantheon, 1968.
215. Jung, C.G. *Psychological Types.* Revised translation by R.F.C. Hull, Vol. 6 of *Collected Works.* Princeton, N.J.: Princeton University Press, 1971.
216. Jung, C.G. *The Structure and Dynamics of the Psyche.* Trans. by R.F.C. Hull. New York: Pantheon, 1960.
217. Kant, Immanuel. *Critique of Practical Reason.* Trans. by Lewis White Beck. New York: Bobbs-Merrill, 1956.
218. Kant, Immanuel. *Critique of Pure Reason.* Trans. by Norman Kemp Smith, London: Macmillan, 1958.
219. Kaplan, Abraham. *The Conduct of Inquiry, Methodology for Behavioral Science.* San Francisco: Chandler, 1964.
220. Kardiner, Abram. "Social and Cultural Implications of Psychoanalysis." In Sidney Hook (Ed.) *Psychoanalysis, Scientific Method, and Philosophy.* New York: New York University Press, 1959.
221. Kash, Don E.; White, Irvin L.; Reuss, John W.; and Leo, Joseph. "University Affiliation and Recognition: National Academy of Sciences." *Science,* March 10, 1972, *175* (4026), 1076—1084.
222. Kaula, William M. "Dynamical Aspects of Lunar Origin." *Reviews of Geophysics and Space Physics,* May 1971, *9* (2),217—238.
223. Kelly, George A. *The Psychology of Personal Constructs,* Vol. 2. New York: Norton, 1955.
224. Keniston, Hayward. *Graduate Study and Research in The Arts and Sciences* at the University of Pennsylvania. Philadelphia, 1958.
225. Kessel, Frank S. "The Philosophy of Science As Proclaimed and Science as Practiced: 'Identity' or 'Dualism?'" *American Psychologist,* Nov. 1969, *24* (11), 999—1005.
226. Kiesler, Charles A. *The Psychology of Commitment, Experiments Linking Behavior to Belief.* New York: Academic Press, 1971.
227. Klahr, David. "A Monte Carlo Investigation of the Statistical Signifi-

cance of Kruskal's Nonmetric Scaling Procedure." *Psychometrika*, Sept. 1969, *34* (3), 319—330.

228. Klopper, Leo E. *Teacher's Guide, Air Pressure.* Chicago: Science Research Associates, 1966.

229. Klopper, Leo E. *Teacher's Guide, The Chemistry of Fixed Air.* Chicago: Science Research Associates, 1964.

230. Klopper, Leo E. *Teacher's Guide, Frogs and Batteries.* Chicago: Science Research Associates, 1966.

231. Kluckhohn, Clyde; and Murray, Henry A. (Eds.) *Personality in Nature, Society, and Culture.* New York: Knopf, 1953.

232. Knapp, R.H. "Demographic, Cultural, and Personality Attributes of Scientists" and "Personality Committee Report." In *Research Conference on the Identification of Creative Scientific Talent.* Salt Lake City: University of Utah Press, 1963.

233. Knower, F.H. "Experimental Studies of Changes in Attitude: II, A Study of the Effect of Printed Argument on Changes in Attitude." *Journal of Abnormal and Social Psychology*, 1936, *30*, 522—532.

234. Koestler, Arthur. *The Sleepwalkers, A History of Man's Changing Vision of the Universe.* New York: Macmillan, 1965.

235. Kohler, Foy D., and Harvey, Dodd L. "Administering and Managing the U.S. and Soviet Space Programs." *Science*, Sept. 11, 1970, *169* (3950), 1049—1056.

236. Kordig, C.R. "The Theory-ladenness of Observation." *Review of Metaphysics*, 1971, *24*, 448—484.

237. Kordig, Carl R. "The Comparability of Scientific Theories." *Philosophy of Science*, Dec. 1971, *38* (4), 467—485.

238. Kornhauser, William. *Scientists in Industry, Conflict and Accommodation.* Berkeley and Los Angeles: University of California Press, 1962.

239. Kosok, Michael. "The Formalization of Hegel's Dialectical Logic." *International Philosophical Quarterly*, 1966, *6*, 596—631.

240. Krikorian, Y.H. "Singer's Philosophy of Experimentalism." *Philosophy of Science*, 1962, *29*, 81—91.

241. Krippendorf, Klaus. "Reliability of Recording Instructions: Multivariate Agreement for Nominal Data." *Behavioral Science*, May 1971, *16* (3), 228—235.

242. Kubie, Lawrence S. *Neurotic Distortion of the Creative Process.* Lawrence, Kansas: University of Kansas Press, 1961.

243. Kubie, Lawrence S. "Some Unsolved Problems of the Scientific Career." *American Scientist*, Oct. 1953, *41*, 596—613; and Jan. 1954, *42*, 104—112.

244. Kuhn, Thomas S. "The Essential Tension: Tradition and Innovation in Scientific Research." In Calvin W. Taylor and Frank Barron (Eds.) *Scientific Creativity, Its Recognition and Development.* New York: Wiley, 1963.

245. Kuhn, Thomas S. "The Function of Measurement in Modern Physical Science." *Isis*, 1961, *52*, 161—193.

246. Kuhn, Thomas S. "Logic of Discovery or Psychology of Research?" In Imre Lakatos and Alan Musgrave (Eds.) *Criticism and the Growth of Knowledge.* Cambridge, England: Cambridge University Press, 1970.

247. Kuhn, Thomas S. *The Structure of Scientific Revolutions.* Chicago: University of Chicago, 1962.
248. Kuhn, Thomas S. *The Structure of Scientific Revolutions.* (2nd ed.) Chicago: University of Chicago, 1970.
249. Ladd, Everett Carll, Jr., and Lipset, Seymour Martin. "Politics of Academic Natural Scientists and Engineers." *Science,* June 9, 1972, *176,* 1091—1100.
250. La Forge, Rolfe, *et al.* "The Interpersonal Dimension of Personality: II, An Objective Study of Repression." *Journal of Personality,* 1954—1955, *23,* 129—153.
251. Lakatos, Imre. "Falsification and the Methodology of Scientific Research Programmes." In Imre Lakatos and Alan Musgrave (Eds.) *Criticism and the Growth of Knowledge.* Cambridge, England: Cambridge University Press, 1970.
252. Lakatos, Imre, and Alan Musgrave (Eds.) *Criticism and the Growth of Knowledge.* Cambridge, England: Cambridge University Press, 1970.
253. Lampkin, Richard H., Jr. "Scientific Attitudes." *Science Education,* Dec. 1938, *22* (7), 353—357.
254. Laudan, Laurens. "Discussion: Grünbaum on 'The Duhemian Argument.' " *Philosophy of Science,* July-Oct. 1965, *32* (3—4), 295—299.
255. Lazarsfeld, Paul F., and Rosenberg, Morris. *The Language of Social Research.* New York: Free Press, 1955.
256. Lazerowitz, Morris. "The Relevance of Psychoanalysis to Philosophy." In Sidney Hook (Ed.) *Psychoanalysis, Scientific Method, and Philosophy.* New York: New York University Press, 1959.
257. Leach, James, "Explanation and Value Neutrality." *British Journal for the Philosophy of Science,* 1968, *19,* 93—108.
258. Leary, Timothy. *Interpersonal Diagnosis of Personality: A Functional Theory and Methodology for Personality Evaluation.* New York: Ronald Press, 1957.
259. Lee, Donald S. "Scientific Method as a Stage Process." *Dialectica,* 1968, *22* (1), 28—44.
260. Levinson, Alfred A. *Proceedings of the Second Lunar Science Conference.* Cambridge, Mass.: Massachusetts Institute of Technology Press, 1971.
261. Levinson, Alfred A., and Taylor, S. Ross. *Moon Rocks and Minerals.* New York: Pergamon, 1971.
262. Lodahl, Janice Beyer, and Gordon, Gerald. "The Structure of Scientific Fields and The Functioning of University Graduate Departments." *American Sociological Review,* Feb. 1972, *37,* 57—72.
263. Logsdon, John M. *The Decision to Go to the Moon, Project Apollo and the National Interest.* Cambridge, Mass.: Massachusetts Institute of Technology Press, 1970.
264. Lorell, J., and Sjogren, W.L. "Lunar Gravity: Preliminary Estimates from Lunar Orbiter." *Science,* Feb. 9, 1968, *159,* 625—627.
265. Luce, R. Duncan, and Raiffa, Howard. *Games and Decisions, Introduction and Critical Survey.* New York: Wiley, 1957.
266. Lumsdaine, Arthur A., and Janis, Irving L. "Resistance to 'Counter-Propaganda' Produced by One-Sided and Two-Sided 'Propaganda' Presentations." *Public Opinion Quarterly,* 1953, *17,* 311—318.

305

267. Lunar Sample Analysis Planning Team. "Third Lunar Science Conference." *Science*, June 2, 1972, *176* (4038), 975—981.
268. Lunar Sample Preliminary Examination Team. "Preliminary Examination of Lunar Samples from Apollo 12." *Science*, March 6, 1970, *167* (3423), 1325—1339.
269. Mahoney, Michael S. "Fermat's Mathematics: Proofs and Conjectures." *Science*, Oct. 6, 1972, *178* (4056), 30—36.
270. Mailer, Norman. *Of a Fire on the Moon.* Boston: Little, Brown, 1970.
271. Mann, Harriet; Siegler, Miriam; and Osmond, Humphry. "Four Types of Personalities and Four Ways of Perceiving Time." *Psychology Today*, Dec. 1972, *6* (7), 76—84.
272. Marcson, Simon. *The Scientist in American Industry, Some Organizational Determinants in Manpower Utilization.* Princeton, N.J.: Princeton University Press, 1960.
273. Marcuse, Herbert. *Eros and Civilization, A Philosophical Inquiry into Freud.* New York: Vintage, 1955.
274. Margolis, Joseph. "Notes on Feyerabend and Hanson." In Michael Radner and Stephen Winokur (Eds.) *Analyses of Theories and Methods of Physics and Psychology, Minnesota Studies in the Philosophy of Science, Vol. IV.* Minneapolis: University of Minnesota Press, 1970, 193—195.
275. Marsden, B.G., and Cameron, A.G.W. *The Earth-Moon System.* New York: Plenum Press, 1966.
276. Martin, Michael. "How To Be A Good Philosopher of Science: A Plea For Empiricism In Matters Methodological." *Boston Studies in The Philosophy of Science*, to appear.
277. Martin, Michael. "Ontological Variance and Scientific Objectivity." *British Journal for the Philosophy of Science*, 1972, *23*, 252—256.
278. Martin, Michael. "Referential Variance and Scientific Objectivity." *British Journal for the Philosophy of Science*, 1971, *22*, 17—26.
279. Martin, Michael. "Theoretical Pluralism." *Philosophia*, to appear.
280. Maslow, Abraham H. *The Psychology of Science.* New York: Harper and Row, 1966.
281. Mason, Brian; and Melson, William G. *The Lunar Rocks.* New York: Wiley-Interscience, 1970.
282. Mason, Richard O. "A Dialectical Approach to Strategic Planning." *Management Science*, April 1969, *15* (8), B-403 to B-414.
283. Matson, Floyd. *The Broken Image, Man, Science, and Society.* New York: Doubleday, 1966.
284. Maxwell, Nicholas. "A Critique of Popper's Views on Scientific Method." *Philosophy of Science*, June 1972, *39* (2), 131—152.
285. McClelland, David C. "On the Dynamics of Creative Physical Scientists." In Liam Hudson (Ed.) *The Ecology of Human Intelligence.* Harmondsworth, England: Penguin, 1970.
286. McClelland, David C. "The Calculated Risk: An Aspect of Scientific Performance." In C.W. Taylor (Ed.) *Research Conference on The Identification of Creative Scientific Talent.* Salt Lake City: University Utah Press, 1956.
287. McCurdy, R.D. "Characteristics of Superior Science Students and Their Own Sub-Groups." *Science Education*, 1956, *40*, 3—24.
288. McNemar, Quinn. *Psychological Statistics.* New York: Wiley, 1962.

289. Medawar, Peter Brian. *Induction and Intuition in Scientific Thought.* Philadelphia: American Philosophical Society, 1969.
290. Medawar, Peter Brian. "Is the Scientific Paper Fraudulent? Yes; It Misrepresents Scientific Thought." *Saturday Review,* August 1, 1964, 42 —43.
291. Merton, Robert K. "The Ambivalence of Scientists." *Bulletin of the Johns Hopkins Hospital,* Feb. 1963, *112* (2), 77—97.
292. Merton, Robert K. "Behavior Patterns of Scientists." *American Scientist,* Spring 1969, *57* (1), 1—23.
293. Merton, Robert K. "The Matthew Effect In Science, The Reward and Communications Systems of Science Are Considered." *Science,* Jan. 5, 1968, *159* (3810), 56—83.
294. Merton, Robert K. "Priorities in Scientific Discovery: A Chapter in the Sociology of Science." In Bernard Barber and Walter Hirsch (Eds.) *The Sociology of Science.* New York: Free Press, 1962.
295. Merton, Robert K. "Resistance to the Systematic Study of Multiple Discoveries in Science." *European Journal of Sociology,* 1963, *4,* 237—282.
296. Merton, Robert K. "Singletons and Multiples in Scientific Discovery: A Chapter in the Sociology of Science." *Proceedings of the American Philosophical Society,* Oct. 1961, *105* (5), 470—486.
297. Merton, Robert K. *Social Theory and Social Structure.* New York: Free Press, 1968.
298. Merton, Robert K., and Barber, Elinor. "Sociological Ambivalence." In E.A. Tiryakian (Ed.) *Sociological Theory, Values, and Sociocultural Change.* New York: Free Press, 1963.
299. Merton, Robert K., and Kendall, Patricia L. "The Focused Interview." *The American Journal of Sociology,* 1946, *51,* 541—557.
300. Merton, Robert K., *et al. The Focused Interview, A Manual of Problems and Procedures.* New York: Free Press, 1956.
301. Miles, W. Martin. "The Measurement of Value of Scientific Information." In Burton V. Dean (Ed.) *Operations Research in Research and Development.* New York: Wiley, 1963.
302. Mills, C. Wright. *The Sociological Imagination.* London: Oxford University Press, 1959.
303. Mitias, R.G.E. "Concepts of Science and Scientists Among College Students." *Journal of Research in Science Teaching,* 1970, 7, 135—140.
304. Mitroff, Ian I. "A Brunswik Lens Model of Dialectical Information Systems." *Theory and Decision,* to appear.
305. Mitroff, Ian I. "A Communication Model of Dialectical Inquiring Systems — A Strategy for Strategic Planning." *Management Science,* June 1971, *17* (10), B-634 to B-648.
306. Mitroff, Ian I. "Epistemology As General Systems Theory: An Approach to the Conceptualization of Complex Decision-Making Experiments." *Philosophy of the Social Sciences,* 1973, *3,* 117—134.
307. Mitroff, Ian I. "Methodological Advances in the Behavioral Sciences and The Value Neutrality Thesis." *Methodology and Science,* 1970, *3* (4), 143—155.
308. Mitroff, Ian I. "The Myth of Objectivity or Why Science Needs a New

Psychology of Science." *Management Science,* June 1972, *18* (10), B-613 to B-618.

309. Mitroff, Ian I. "The Mythology of Methodology: An Essay on The Nature of a Feeling Science." *Theory and Decision,* March 1972, *2* (3), 274—290.

310. Mitroff, Ian I. "On the Methodology of the Holistic Experiment: An Approach to the Conceptualization of Large-Scale Social Experiments." *Journal of Technological Forecasting and Social Change,* 1973, *4,* 339—353.

311. Mitroff, Ian I. "On The Social Psychology of The Safety Factor: A Case Study in The Sociology of Engineering Science." *Management Science,* April 1972, *18* (8), B-454 to B-469.

312. Mitroff, Ian I. "Simulating Engineering Design: A Case Study on the Interface Between The Technology and Social Psychology of Design." *IRE Transactions on Engineering Management.* Dec. 1968, *EM-15* (4), 178—187.

313. Mitroff, Ian I. "Solipsism: An Essay in Psychological Philosophy." *Philosophy of Science,* Sept. 1971, *38* (3), 376—394.

314. Mitroff, Ian I. "Systems, Inquiry, and The Meanings of Falsification." *Philosophy of Science,* 1973, *40* (2), 255—276.

315. Mitroff, Ian I., and Betz, Frederick. "Dialectical Decision Theory: A Meta-Theory of Decision-Making." *Management Science,* Sept. 1972, *19* (1), 11—24.

316. Mitroff, Ian I.; Nelson, John; and Mason, Richard O. "On the Theory and Nature of Anecdoctal Information Systems." *Management Science,* to appear.

317. Mogar, Robert E. "Toward a Psychological Theory of Education." *Journal of Humanistic Psychology,* Spring 1969, *9* (1), 17—52.

318. Morgenbesser, Sidney. "Psychologism and Methodological Individualism." In Sidney Morgenbesser (Ed.) *Philosophy of Science Today,* New York: Basic Books, 1967.

319. Morris, William T. "Intuition and Relevance." *Management Science,* Dec. 1967, *14* (4), B-157 to B-165.

320. Muller, P.M., and Sjogren, W.L. "Mascons: Lunar Mass Concentrations." *Science,* Aug. 16, 1968, *161,* 680—684.

321. Mumford, Lewis. "Reflections: The Megamachine — I." *The New Yorker,* Oct. 10, 1970, 50—131.

322. Mumford, Lewis. "Reflections: The Megamachine — II." *The New Yorker,* Oct. 17, 1970, 48—141.

323. Mumford, Lewis. "Reflections: The Megamachine — III." *The New Yorker,* Oct. 24, 1970, 55—127.

324. Mumford, Lewis. "Reflections: The Megamachine — IV." *The New Yorker,* Oct. 31, 1970, 50—98.

325. Murray, Henry A. *Explorations In Personality.* New York: Oxford, 1938.

326. Murray, Henry A. (Ed.) *Myth and Mythmaking.* New York, George Braziller, 1960.

327. Myers-Briggs, I. *Manual for the Myers-Briggs Type Indicator.* Princeton, N.J.: Educational Testing Service, 1962.

328. Myrdal, Gunnar. *An American Dilemma: The Negro Problem and Modern Democracy*, Vol. 2. New York: Harper and Row, 1962.
329. Nagel, Ernest. "Methodological Issues in Psychoanalytic Theory." In Sidney Hook (Ed.) *Psychoanalysis, Scientific Method, and Philosophy.* New York: New York University Press, 1959.
330. Nagel, Ernest. *The Structure of Science, Problems in the Logic of Scientific Explanation.* New York: Harcourt, Brace, and Jovanovich, 1961.
331. Nelson, Benjamin. "The Early Modern Revolution in Science and Philosophy." In Robert S. Cohen and Marx W. Wartofsky (Eds.) *Boston Studies in the Philosophy of Science, Vol. III.* Dordrecht, Holland: D. Reidel, 1967.
332. Nidditch, P.H. (Ed.) *The Philosophy of Science.* London: Oxford University Press, 1968.
333. Nisbet, Robert A. "Sociology as an Art Form." In Maurice Stein and Arthur Vidich (Eds.) *Sociology on Trial.* Englewood Cliffs, N.J.: Prentice-Hall, 1963.
334. Nunnally, Jum. "The Analysis of Profile Data." *Psychological Bulletin,* 1962, *59* (4), 311—319.
335. O'Keefe, John A. (Ed.) *Tektites.* Chicago: University of Chicago, 1963.
336. Orne, Martin T. "On The Social Psychology of the Psychological Experiment: With Particular Reference to Demand Characteristics and Their Implications." *American Psychologist,* 1962, *17,* 776—783.
337. Osgood, Charles E. "Semantic Space Revisited." In James Snider and Charles E. Osgood (Eds.) *Semantic Differential Technique, A Sourcebook.* Chicago: Aldine, 1969.
338. Osgood, Charles E.; Suci, George J.; and Tannenbaum, Percy H. *The Measurement of Meaning.* Urbana, Ill.: University of Illinois, 1957.
339. Patrick, Catharine. "Scientific Thought." *Journal of Psychology,* 1938, *5,* 55—83.
340. Payne, Stanley L. *The Art of Asking Questions.* Princeton, N.J.: Princeton University Press, 1951.
341. Planck, Max. *Scientific Autobiography and Other Papers.* Trans. by Frank Gaynor. New York: Philosophical Library, 1949.
342. Polanyi, Michael. "Logic and Psychology." *American Psychologist,* 1968, *23,* 27—43.
343. Polanyi, Michael. *Personal Knowledge, Towards a Post-Critical Philosophy.* New York: Harper, 1964.
344. Polanyi, Michael. *Science, Faith, and Society.* Chicago: University of Chicago Press, 1964.
345. Polanyi, Michael. *The Tacit Dimension,* New York: Doubleday, 1967.
346. Popper, Karl R. *Conjectures and Refutations, The Growth of Scientific Knowledge.* New York: Basic Books, 1962.
347. Popper, Karl R. *The Logic of Scientific Discovery.* New York: Harper and Row, 1965.
348. Popper, Karl R. "Normal Science and Its Dangers." In Imre Lakatos and Alan Musgrave (Eds.) *Criticism and the Growth of Knowledge.* New York: Cambridge University Press, 1970.
349. Popper, Karl R. *The Poverty of Historicism.* New York: Harper and Row, 1961.

350. Price, Derek J. de Solla. *Little Science, Big Science.* New York: Columbia University Press, 1963.
351. Price, Derek J. de Solla. "The Science of Science." *Bulletin of the Atomic Scientists,* Oct. 1965, 2—8.
352. Price, Don K. *The Scientific Estate.* Cambridge, Mass.: Belknap, 1965.
353. Quinn, Philip L. "The Status of the D-Thesis." *Philosophy of Science,* Dec. 1969, *36* (4), 381—399.
354. Radner, Michael, and Winokur, Stephen (Eds.) *Analyses of Theories and Methods of Physics and Psychology, Vol. IV, Minnesota Studies in the Philosophy of Science.* Minneapolis: University of Minnesota Press, 1970.
355. Radnitzky, Gerard. "Ways of Looking at Science: A Synoptic Study of Contemporary Schools of 'Metascience.'" *General Systems,* 1969, *14,* 187—191.
356. Ravetz, Jerry. "Ideological Crisis in Science." *New Scientist and Science Journal,* July 1, 1971, *51* (758), 35—36.
357. Ravetz, Jerry. *Scientific Knowledge and Its Social Problems.* Oxford, England: Clarendon Press, 1971.
358. Reichenbach, Hans. *Experience and Prediction, An Analysis of the Foundations and the Structure of Knowledge.* Chicago: University of Chicago Press, 1938.
359. Reichenbach, Hans. *The Rise of Scientific Philosophy.* Berkeley and Los Angeles: University of California Press, 1968.
360. Reif, F. "The Competitive World of the Pure Scientist." In Norman Kaplan (Ed.) *Science and Society.* Chicago: Rand McNally, 1965.
361. Rescher, Nicholas. *The Coherence Theory of Truth.* Oxford, England: Clarendon Press, 1973.
362. Rescher, Nicholas. "The Ethical Dimension of Scientific Research." In Robert G. Colodny (Ed.) *Beyond the Edge of Certainty.* Englewood Cliffs, N.J.: Prentice-Hall, 1965.
363. Richardson, Stephen A.; Dohrenwend, Barbara Snell; and Klein, David. *Interviewing, Its Forms and Functions.* New York: Basic Books, 1965.
364. Roberts, Walter Orr. "After The Moon, The Earth!" *Science,* Jan. 2, 1970, *167* (3914), 11—16.
365. Roe, Anne. "Changes in Scientific Activities with Age." *Science,* Oct. 15, 1965, *150,* 313—318.
366. Roe, Anne. *The Making of A Scientist.* New York: Dodd, Mead, 1953.
367. Roe, Anne. "A Psychological Study of Eminent Physical Scientists." *Genetic Psychology Monographs,* 1951, *43,* 121—235.
368. Roe, Anne. "A Psychological Study of Eminent Psychologists and Anthropologists, and a Comparison with Biological and Physical Scientists." *Psychological Monograph,* 1953, *67* (2).
369. Roe, Anne. "A Psychologist Examines 64 Eminent Scientists." *Scientific American,* Nov. 1952, *187* (5), 21—25.
370. Roe, Anne. *The Psychology of Occupations.* New York: Wiley, 1956.
371. Roe, Anne. "The Psychology of The Scientist." *Science,* Aug. 18, 1961, *134,* 456—459.
372. Roe, Anne. "Scientists Revisited." *Harvard Studies in Career Development.* No. 38. Cambridge, Mass.: Graduate School of Education, Harvard University, 1965.

373. Roe, Anne. "Women in Science." *Personnel and Guidance Journal,* April 1966, *44* (8), 784—787.

374. Roe, Anne; and Mierzwa, John. "The Use of The Rorschach in The Study of Personality and Occupations." *Journal of Projective Techniques,* Sept. 1960, *24* (3), 282—289.

375. Rokeach, Milton. *Beliefs, Attitudes and Values.* San Francisco: Jossey-Bass, 1968.

376. Rokeach, Milton. *The Open and Closed Mind, Investigations Into the Nature of Belief Systems and Personality Systems.* New York: Basic Books, 1960.

377. Roose, Kenneth C., and Andersen, Charles J. *A Rating of Graduate Programs.* Washington, D.C.: American Council on Education, 1970.

378. Rose, Arnold. "The Relation of Theory and Method." In Llewellyn Gross (Ed.) *Sociological Theory.* New York, Harper and Row, 1967.

379. Rosenthal, Robert. *Experimenter Effects in Behavioral Research.* New York: Appleton-Century Crofts, 1966.

380. Roszak, Theodore. *The Making of A Counter Culture, Reflections on the Technocratic Society and Its Youthful Opposition.* New York: Anchor, 1969.

381. Roszak, Theodore. "Science: A Technocratic Trap." *The Atlantic,* July 1972, *230* (1), 56—61.

382. Roszak, Theodore. *Where The Wasteland Ends, Politics and Transcendence in Postindustrial Society.* New York: Doubleday, 1972.

383. Rubenstein, Albert H., and Sullivan, Edward M. (Eds.) A *Directory of Research-On-Research.* Evanston, Ill.: Technological Institute, Northwestern University, 1968.

384. Rudner, Richard S. *Philosophy of Social Science.* Englewood Cliffs, N.J.: Prentice-Hall, 1966.

385. Rummel, R.J. "Understanding Factor Analysis." *Journal of Conflict Resolution,* Dec. 1967, *11* (4), 444—480.

386. Runcorn, S.K. "Lunar Dust." *Science Journal,* May 1970, *6* (5), 27—32.

387. Sarton, George. *The History of Science and the New Humanism.* Bloomington, Ind.: Indiana University Press, 1962.

388. Sarton, George. *The Life of Science, Essays in the History of Civilization.* Bloomington, Ind.: Indiana University Press, 1960.

389. Sarton, George. *The Study of the History of Mathematics, and the Study of the History of Science.* New York: Dover, 1957.

390. Scheffler, Israel. "Discussion: Vision and Revolution: A Postscript On Kuhn." *Philosophy of Science,* Sept. 1972, *39.* (3), 366—374.

391. Scheffler, Israel. *Science and Subjectivity.* New York: Bobbs-Merrill, 1967.

392. Schuessler, Karl. *Analyzing Social Data, A Statistical Orientation.* Boston: Houghton Mifflin, 1971.

393. Schwab, Joseph J. "What Do Scientists Do?" *Behavioral Science,* 1960, *5* (1), 1—27.

394. Settle, Tom. "The Rationality of Science *versus* The Rationality of Magic." *Philosophy of the Social Sciences,* Sept. 1971, *1* (3), 173—194.

395. Shannon, Claude E., and Weaver, Warren. *The Mathematical Theory of Communication.* Urbana, Ill.: University of Illinois Press, 1964.

396. Shapere, Dudley. "Meaning and Scientific Change." In Robert G. Colod-

ny (ed.) *Mind and Cosmos*. Pittsburgh: University of Pittsburgh Press, 1966.

397. Shapere, Dudley. "The Paradigm Concept: A Review of 'The Structure of Scientific Revolutions' by Thomas S. Kuhn and 'Criticism and The Growth of Knowledge' by Imre Lakatos and Alan Musgrave (Eds.)" *Science*, May 14, 1971, *172* (3984), 706—709.

398. Shapere, Dudley. *Philosophical Problems of Natural Science*. London: Macmillan, 1965.

399. Shapley, Deborah. "Nobelists: Piccioni Lawsuit Raises Questions About the 1959 Prize." *Science*, June 30, 1972, *176*, 1405—1406.

400. Shen, Eugene. "The Validity of Self-Estimate." *Journal of Educational Psychology*, 1925, *16*, 104—107.

401. Shenk, Charlotte F. *Tektites: A Bibliography*. Redstone Arsenal, Alabama: Research Branch, Redstone Scientific Information Center, U.S. Army, January 15, 1964.

402. Sherman, Mandel. "The Interpretation of Emotional Responses in Infants." In Wayne Dennis (Ed.) *Readings in Child Psychology*. Englewood Cliffs, N.J.: Prentice-Hall, 1951.

403. Shibutani, Tamotsu. "Reference Groups as Perspectives." *American Journal of Sociology*, 1955, *60*, 562—569.

404. Short, Nicholas M. *Planetary Geoscience*. Council on Education in The Geological Sciences, Englewood Cliffs, N.J.: Prentice-Hall, 1971.

405. Silverman, Irwin; Shulman, Arthur D.; and Wiesenthal, David L. "Effects of Deceiving and Debriefing Psychological Subjects on Performance in Later Experiments." *Journal of Personality and Social Psychology*, 1970, *14* (3), 203—212.

406. Simon, Herbert A. *The Sciences of the Artificial*. Cambridge, Mass.: MIT Press, 1969.

407. Singer, Barry F. "Toward A Psychology of Science." *The American Psychologist 1971*, 26, *1010—1015*.

408. Singer, Edgar A. In C. West Churchman (Ed.) *Experience and Reflection*. Philadelphia: University of Pennsylvania Press, 1959.

409. Singer, Edgar A. "Mechanism, Vitalism, Naturalism, A Logico-Historical Study." *Philosophy of Science*, April 1946, *13* (2), 81—99.

410. Singer, Edgar A. *Mind as Behavior and Studies in Empirical Idealism*. Columbus, O.: R.G. Adams, 1924.

411. Singer, Edgar A. *Modern Thinkers and Present Problems; An Approach to Modern Philosophy Through Its History*. New York: Holt, 1923.

412. Singer, Edgar A. "On the Conscious Mind." *The Journal of Philosophy*, Oct. 1929, *26* (21), 561—575.

413. Sjoberg, Gideon and Nett, Roger. *A Methodology for Social Research*. New York: Harper and Row, 1968.

414. Smith, H.C. "Sensitivity to People." In Hans Toch and Clay Smith (Eds.) *Social Perception*, New York: Van Nostrand, 1968.

415. Smith, Henry Bradford. "Postulates of Empirical Thought." *Journal of Philosophy*, June 7, 1928, *25* (12), 318—323.

416. Snider, James G., and Osgood, Charles E. *Semantic Differential Technique, A Sourcebook*. Chicago: Aldine, 1969.

417. Spranger, Edward. *Types of Men, The Psychology and Ethics of Personality*. Halle: Niemeyer, 1928.

418. Staats, Arthur W., and Staats, Carolyn K. "Meaning and *m*, Correlated but Separate." In James G. Snider and Charles E. Osgood (Eds.) *Semantic Differential Technique, A Sourcebook.* Chicago: Aldine, 1969.

419. Staub, Ervin. "Instigation to Goodness: The Role of Social Norms and Interpersonal Influence." *Journal of Social Issues,* 1972, *28* (3), 131–150.

420. Stein, M.I. "A Transactional Approach to Creativity." In C.W. Taylor (Ed.) *Research Conference on the Identification of Creative Scientists.* Salt Lake City: University of Utah Press, 1956.

421. Stein, Maurice R. "The Poetic Metaphors of Sociology." In Maurice Stein and Arthur Vidich (Eds.) *Sociology on Trial.* Englewood Cliffs, N.J.: Prentice-Hall, 1963.

422. Stevens, S.S. "The Operational Basis of Psychology." *American Journal of Psychology,* 1935, *47,* 323–330.

423. Storer, Norman W. *The Social System of Science.* New York: Holt, Rinehart, and Winston, 1966.

424. Suits, Bernard. "Is Life A Game We Are Playing?" *Ethics,* April 1967, *77* (3), 209–213.

425. Suits, Bernard. "What Is A Game?" *Philosophy of Science.* 1967, *34,* 148–156.

426. Suppe, Frederick. "What's Wrong With The Received View On the Structure Of Scientific Theories?" *Philosophy of Science,* March 1972, *39* (1), 1–19.

427. Swatez, Gerald M. "Social Organization of A University Laboratory." Unpublished doctoral dissertation. Internal Working Paper No. 44, Space Sciences Laboratory, Social Sciences Project, University of California, Berkeley, April 1966.

428. Swatez, Gerald M. "Some Social Characteristics of The Research Process in High-Energy Physics: A Case Study of a Large Hydrogen Bubble-Chamber Group." Internal Working Paper No. 22, Space Sciences Laboratory, Social Sciences Project, University of California, Berkeley, December, 1964.

429. Szent-Gyorgyi, Albert. "Dionysians and Apollonians." Letter to *Science,* June 2, 1972, *176* (4038), 966.

430. Tagiuri, Renato. "Value Orientations and the Relationship of Managers and Scientists." *Administrative Science Quarterly,* June, 1965, *10* (1), 39–51.

431. Takeuchi, H.; Uyeda, S.; and Kanamori, H. *Debate About The Earth: Approach To Geophysics Through Analysis Of Continental Drift.* Trans. Keiko Kanamoi. San Francisco: Freeman, Cooper, 1967.

432. Tart, Charles T. "States of Consciousness and State-Specific Sciences." *Science,* June 16, 1972, *176* (4040), 1203–1210.

433. Taylor, A.M. *Imagination and the Growth of Science.* New York: Schocken, 1970.

434. Taylor, Calvin W., and Barron, Frank (Eds.) *Scientific Creativity: Its Recognition and Development.* New York: Wiley, 1963.

435. Teevan, Richard C. "Personality Correlates of Undergraduate Field of Specialization." *Journal of Consulting Psychology,* 1954, *18,* 212–214.

436. Terman, Lewis M. "Scientists and Nonscientists in a Group of 800 Gifted Men." *Psychological Monograph,* 1954, *68* (7).

437. Theil, Henri. "On The Use of Information Theory Concepts in The Analysis of Financial Statements." *Management Science*, May 1969, *15* (9), 459—480.

438. Thomson, George. "The Two Aspects of Science." *Science*, Oct. 14, 1960, *132*, 996—1000.

439. Thomson, J.J. *The Corpuscular Theory of Matter*. London: Archibald Constable, 1907.

440. Thurstone, L.L. "A Multiple Factor Study of Vocational Interests." *Personnel Journal*, 1931, *10*, 198—205.

441. Tobey, Ronald C. *The American Ideology of National Science, 1919—1930*. Pittsburgh: University of Pittsburgh, 1971.

442. Toch, Hans; and Smith, Henry Clay (Eds.) *Social Perception, The Development Of Interpersonal Impressions, An Enduring Problem in Psychology*. Princeton, N.J.: Van Nostrand, 1968.

443. Torgerson, Warren S. "Multidimensional Scaling: I. Theory and Method." *Psychometrika*, 1952, *17* (4), 401—418.

444. Torgerson, Warren S. "Multidimensional Scaling of Similarity." *Psychometrika*, 1965, *30* (4), 379—393.

445. Torgerson, Warren S. *Theory and Methods of Scaling*. New York: Wiley, 1958.

446. Toulmin, Stephen. *Human Understanding, Vol. I: The Collective Use and Evolution of Concepts*. Princeton, N.J.: Princeton University Press, 1972.

447. Turner, Merle B. *Psychology and The Philosophy of Science*. New York: Appleton—Century-Crofts, 1968.

448. Turoff, Murray. "Delphi Conferencing." Technical Memorandum TM-125, United States Office of the Assistant Director for Resource Analysis, March 1971.

449. Tyler, Forrest B. "Shaping of the Science." *American Psychologist*, March 1970, *25* (3), 219—226.

450. Vas Dias, Robert. *Inside Outer Space: New Poems of the Space Age*. New York: Doubleday, 1970.

451. Verne, Jules. *A Journey to the Center of the Earth*. New York: Scholastic Book Services, 1965.

452. Vernon, M.C. "The Relationship of Occupation to Personality." *British Journal Of Psychology*, April 1941, *31* (4), 297—326.

453. Vernon, P.E. "The Assessment of Psychological Qualities by Verbal Methods." *Industrial Health Research Board Reports*, No. 83, H.M. Statistical Office, London, 1938.

454. Wade, Nicholas. "Theodore Roszak: Visionary Critic of Science." *Science*, Dec. 1, 1972, *178* (4064), 960—962.

455. Wakita, Hiroshi; and Schmitt, Roman A. "Lunar Anorthosites: Rare-Earth and Other Elemental Abundances." *Science*, Nov. 27, 1970, *170* (3961), 969—974.

456. Walbert, Herbert J. "A Portrait of the Artist and Scientist as Young Men." *Exceptional Children*, Sept. 1969, *36* (1), 5—11.

457. Wartofsky, Marx. "Metaphysics as Heuristic for Science." In Robert S. Cohen and Marx W. Wartofsky (Eds.) *Boston Studies in the Philosophy of Science*. Vol. III. Dordrecht, Holland: D. Reidel, 1967.

458. Watson, James D. *The Double Helix, A Personal Account of the Discovery of the Structure of DNA.* New York: Atheneum, 1968.

459. Webb, Eugene J.; Campbell, Donald T.; Schwarz, Richard D.; and Sechrest, Lee. *Unobtrusive Measures: Non-reactive Research in The Social Sciences.* Chicago: Rand McNally, 1966.

460. Weber, Max. "Science as a Vocation." In H.H. Gerth and C. Wright Mills (Trans. and Eds.) *From Max Weber: Essays in Sociology.* New York: Oxford University Press, 1946.

461. Wedeking, Gary. "Duhem, Quinn, and Grünbaum on Falsification." *Philosophy of Science.* Dec. 1969, *36* (4), 375—380.

462. Weinberg, Alvin M. "The Axiology of Science." *American Scientist,* Nov.-Dec. 1970, *58*, 612—617.

463. Weinberg, Alvin M. "In Defense of Science." *Science,* Jan. 9, 1970, *167* (3915), 141—145.

464. Weinberg, Alvin M. *Reflections on Big Science.* Cambridge, Mass.: MIT Press, 1967.

465. Weinreich, Uriel. "A Rejoinder to Semantic Space Revisited." In James G. Snider and Charles E. Osgood (Eds.) *Semantic Differential Technique, A Sourcebook.* Chicago: Aldine, 1969.

466. Weinreich, Uriel. "Travels Through Semantic Space." In James G. Snider and Charles E. Osgood (Eds.) *Semantic Differential Technique, A Sourcebook.* Chicago, Aldine, 1969.

467. Weisskopf, Victor F. "The Significance of Science." *Science,* April 14, 1972, *176* (4031), 138—146.

468. West, S.S. "The Ideology of Academic Scientists." *Institute of Radio Engineers Transactions on Engineering Management,* June 1972, *EM-7* (2), 54—62.

469. Wetherill, George W. "Of Time and the Moon." *Science,* July 30, 1971, *173* (3995), 383—392.

470. Westfall, Richard S. "Newton and the Fudge Factor." *Science,* Feb. 23, 1973, *179*, 751—756.

471. Weyrauch, Walter O. *The Personality of Lawyers.* New Haven: Yale University Press, 1964.

472. Whipple, Fred L. *Earth, Moon, and Planets.* (3rd ed.) Cambridge, Mass.: Harvard University Press, 1968.

473. White, Martha S. "Psychological and Social Barriers to Women in Science." *Science,* Oct. 23, 1970, *170* (3956), 413—416.

474. Wick, Gerald. "They Won't Shut Up." *New Scientist and Science Journal,* May 27, 1971, 532—533.

475. Wilder, R.L. "The Role of Intuition." *Science,* May 5, 1967, *156*, 605—610.

476. Wilson, J. Tuzo (Ed.) *Continents Adrift, Readings from "Scientific American."* San Francisco: W.H. Freeman, 1972.

477. Wilson, Mitchell. "How Nobel Prizewinners Get That Way." *The Atlantic,* Dec. 1969, *224* (6), 69—74.

478. Wilson, Mitchell. *Passion To Know, The World's Scientists.* New York: Doubleday, 1972.

479. Wisdom, J.O. "Science versus the Scientific Revolution." *Philosophy of The Social Sciences,* May 1971, *1* (2), 123—144.

480. Witkin, H.A., *et al. Psychological Differentiation: Studies of Development.* New York: Wiley, 1962.
481. Wolman, Benjamin B. "Does Psychology Need Its Own Philosophy of Science?" *American Psychologist*, 1971, *26*, 877—886.
482. Wood, John A. "The Lunar Soil." *Scientific American*, Aug. 1970, *223* (2), 14—23.
483. Wylie, Ruth C. *The Self Concept: A Critical Survey of Pertinent Research Literature.* Lincoln. Neb.: University of Nebraska, 1961.
484. Young, Hugo; Silcock, Bryan; and Dunn, Peter. *Journey to Tranquility.* New York: Doubleday, 1970.
485. Young, Hugo; Silcock, Bryan; and Dunn, Peter. "Why We Went to the Moon: From the Bay of Pigs to the Sea of Tranquility." *The Washington Monthly*, April 1970, *2* (2), 28—58.
486. Ziman, J.M. *Public Knowledge, An Essay Concerning the Social Dimensions of Science.* New York: Cambridge University Press, 1968.
487. Zuckerman, Harriet A. "Patterns of Name Ordering Among Authors of Scientific Papers: A Study of Social Symbolism and Its Ambiguity." *American Journal of Sociology*, Nov. 1968, *74* (3), 276—291.

Acknowledgements

I wish to thank the following publishers and authors for permission to quote from their works:

Aldine Publishing Co.--for passages from Russell L. Ackoff and Fred Emery, *On Purposeful Systems* (Chicago: Aldine-Atherton, 1972). David C. McClelland, "On the Dynamics of Creative Physical Scientists," ed. Liam Hudson (Chicago: Aldine-Atherton, 1970).

Basic Books, Inc.--for passages from C. West Churchman, *The Design of Inquirying Systems: Basic Concepts in Systems Analysis*, Copyright 1971, Basic Books (New York); Norwood Russell Hanson, "Observation and Interpretation," *Philosophy of Science Today*, ed. Sidney Morgenbesser, Copyright, 1967, by Basic Books (New York).

Bobbs-Merrill Co., Inc.--for passages from Immanuel Kant, *Critique of Practical Reason*, trans. Lewis White Beck. Copyright, 1956, Liberal Arts Press Inc., reprinted by permission of The Bobbs-Merrill, Co., Inc.

Cambridge University Press--for passages from John Ziman, *Public Knowledge, An Essay Concerning the Social Dimension of Science* (Cambridge, England: Cambridge University Press, 1968).

The University of Chicago Press--for passages from Bernard Suits, "Is Life a Game We Are Playing?" *Ethics*, April 1967, 77, 209—213.

Professor C. West Churchman, School of Business Administration, University of California, Berkeley--for passages from C. West Churchman and Russell L. Ackoff, *Methods of Inquiry, An Introduction to Philosophy and Scientific Method* (Saint Louis: Educational Publishers, 1950).

University of Illinois Press--for passages from Charles E. Osgood and others, *The Measurement of Meaning* (Urbana: University of Illinois Press, 1957).

Institute of Radio Engineers Transactions on Engineering Management--for passages from S.S. West, "The Ideology of Academic Scientists," 1960, *EM-7*, 54-62.

The Johns Hopkins University Press--for passages from Robert K. Merton "The Ambivalence of Scientists," 1963, *112*, 77-97. Copyright 1963, The Johns Hopkins University Press (Baltimore).

University Press of Kansas--for passages from Lawrence Kubie, *Neurotic Distortion of the Creative Process* (Lawrence: University Press of Kansas, 1961).

Little, Brown, and Company Pubs.--for passages from Norman Mailer, *Of a Fire on the Moon* (Boston, Mass.: Little, Brown, 1970).

Macmillan Publishing Co.--for passages from Robert K. Merton and Elinor Barber, "Sociological Ambivalence," *Sociological Theory, Values, and*

317

1065-1075; and "Paradoxes of Science Administration," September 15, 1972, *177*, 964-966; Bentley Glass, "The Ethical Basis of Science," December 3, 1965, *150*, 1254-1261; Albert Szent-Gyorgyi, "Dionysians and Apollonians," June, 1972, *176*, 966.

The journal *Science Education*--for passages from Robert L. Ebel, "What Is the Scientific Attitude?" 1938, *22*, 1-5, 75-81; Richard H. Lampkin, "Scientific Attitudes," 1938, *22*, 353—357.

The journal *Science Teacher*--for passages from Paul B. Diederich, "Components of the Scientific Attitude," 1967, *34*.

Wesleyan University Press--for passages from Norman O. Brown, *Life Against Death, The Psychoanalytical Meaning of History*. Copyright, 1959, by Wesleyan University Press (Middletown, Conn.).

Index

324

Reichenbach distinction, 22, 23, 74.
  See also Discovery and justification,
  contexts of
Relativism, 259, 268, 290
Rescher, Nicholas, 276
Research, Styles of. *See* Cognitive
  styles
Roche limit, 51
Roe, Anne, 141, 142
Rogerian nondirective interviewing, 43
Rose, Arnold, 229
Rosenthal, Robert, 243
Roszak, Theodore, 290

Sample
  formation of, 33—35
  size of, 30—33, 269
Sample scientists
  ages of, 35—36
  aggressiveness and, 144—145
  disciplines of, 36—37
  educational level of, 35
  eminence of, 35
  emotional reactions of, 100, 209
  institutional affiliations of, 31—32
  lunar sample access of, 32—33
  nationalities of, 36
  number of, 31—32
  religious affiliations of, 283
  selection of, 33—34
  typology of, 90—96
Sample's opinions
  of Apollo results, 191—193
  of Apollo scientists, 191
  of causality, 185—188 *passim*
  of competition, 68—70
  of conservatism, 279—280
  of commitment and bias, 64—72,
    162—165, 208
  of geology, 88
  on going to the moon, 202—204
  on honesty, 74
  on Ideal Scientist, 107—131
  on Individual Scientists as types,
    131—137, 162—165, 173—178
  of impartiality, 113—114
  of intuition, 65
  of manned flights, 204
  measurement of changes in, 151—
    160

of metaphysical issues, 186—187 ˙
methods of analysis of, 138—139
of moon as a cultural object, 206—
  208
of NASA, 197, 200—201
of objectivity, 64—65, 97
of politics, 126—127, 277, 278, 279
of scientific behavior, 98—99
on scientific method, 184—185
on scientific speculation, 85—89
  passim
on scientist-astronaut, 200, 205—207
on selection of sample, 35
on social problems and Apollo, 205
on stealing in science, 74—75, 122
of Storybook version of science, 64
strength of, 90, 209
of study, 98, 212—216
on theories of lunar origin, 55—61
on theory vs. data, 183—185
on types of scientists, 85—89, 93—97
Sarton, George, 273
Saturday Review, 288
Saturn, 51
Scheffler, Israel, 285, 287
Schema. *See* Inquiring systems (IS)
Science
  aggressiveness and, 264
  Churchmanian view of, 22—23
  consensus in, 226, 227, 228
  emotional neutrality in, 8—10
  emotions in, 99—100
  esthetics and, 266—267, 268, 287
  ethics and, 233—234, 266—267, 287
  feeling (Jungian) in, 190, 283—284
  feminity in, 210, 211, 264—265
  as a game, 252—255
  interview questions on, 43—45
  intuition in, 97, 268
  lunar, 6
  masculinity of, 112, 130—131, 139,
    142, 144—145, 210—211, 264—
    265
  objectives of, 265—268
  politics in, 124—125, 277, 278, 279
  secrecy in, 75—76
  and social science, 146—147
  social psychology of, 19—22
  sociology and, 10—13
  stealing in, 74—77, 122, 276—277

Synthetic multimodel systems, 228
Systems of Inquiry. *See* Inquiring
    Systems (IS)
Systems separability, 285
Szent-Gyorgyi, Albert, quoted, 211–
    212

TAT (Thematic Apperception Test),
    141, 142, 144, 209
Taylor, A.M., 273; quoted, 18
Taylor, S. Ross, 50; quoted, 36
Tektites, origin of
    theories of, 149–150
    changes in opinion on, 152–153,
        154, 155–156
Terman, Lewis M., 141, 142
Theoretical Language (T.L.), 225
Theoretical Monism (T.M.), 262–263
Theory of science, 247
Theory testing, 239, 240
Theory vs. data, 239, 240, 286, 287
    attitudinal items on, 180
    sample's opinions of, 183–185
Thermal history, lunar
    changes in opinions of, 152–154
    theories of, 148–149, 151
Thomas, W.I., quoted, 78

Thompson, J.J., quoted, 19
Thurstone, L.L., 170
Toulmin, Stephen, 22, 261, 290;
    quoted, 251, 255, 258, 263–264
Truth nets, 222
Typology, Jungian. *See* Jungian
    typology
Typology of scientists, 83–84, 85–89,
    195, 284

USGS (U.S. Geological Survey), 31

Vas Dias, Robert, quoted, 4
Verne, Jules, quoted, 49

Walberg, Herbert J., 140
Wartofsky, Marx, 284
Watson, James D., 70, 98
Webb, Eugene J., 41
Weinreich, Uriel, 105
West, S.S., 276
Weyrauch, Walter O., 282
"What is a Game?", 251
Women's Mysteries Ancient and Modern,
    207

Ziman, John, quoted, 3, 226